Construction Business for the 21st Century

Construction Business for the 21st Century: Managing Risk and Improving Profits covers the business side of construction contracting. This book provides 21st-century processes, procedures, and practices that will advance the construction industry into the future.

The industry that built the United States suffers from diminishing profit margins, causing it to have the second highest failure rate in the nation. Drawing from decades of research into the causes of contractor failure, the author offers solutions for improved profits and risk management and identifies "legacy" industry practices that have become obsolete. The topics covered, such as project selection, the low-bid process, risk identification and management, and understanding market cycles, are critical subjects for 21st-century construction professionals. This book also addresses in detail the important subjects of succession planning, industry consolidation, growth limits, joint ventures, equipment management, and boards of directors/advisors. Readers will receive critical insights into financial and accounting processes, which are rarely covered in technicians' and engineers' education.

Whether a general contractor, construction manager, specialty contractor, credit grantor, or student, this book will provide a thorough understanding of the intricacies and complexities needed to succeed in the construction business.

Thomas C. Schleifer entered the construction industry at age 16 and has more than 60 years of contracting and consulting experience—from Field Superintendent, to contractor, to nationally known consultant, author, and lecturer. His blog and newsletter, Contractor's Weekly Construction Message, has an audience of over 20,000 readers. His previous books include *The Secrets to Construction Business Success* (Routledge 2022).

Construction Business for the 21st Century

Managing Risk and Improving Profits

Thomas C. Schleifer

Routledge
Taylor & Francis Group
NEW YORK AND LONDON

Designed cover image: Shutterstock

First published 2026
by Routledge
605 Third Avenue, New York, NY 10158

and by Routledge
4 Park Square, Milton Park, Abingdon, Oxon, OX14 4RN

Routledge is an imprint of the Taylor & Francis Group, an informa business

ISBN: 978-1-032-81539-8 (hbk)
ISBN: 978-1-032-81523-7 (pbk)
ISBN: 978-1-003-50031-5 (ebk)

DOI: 10.1201/9781003500315

Typeset in Times New Roman
by KnowledgeWorks Global Ltd.

Contents

About the Author

THOMAS C. SCHLEIFER, Ph.D.

Dr. Schleifer joined the construction industry at age 16 and brings more than 60 years of contracting and consulting experience to his presentations. He has Bachelor and Master of Science degrees in construction management from East Carolina University and a Ph.D. in construction management from Heriot-Watt University, Edinburgh, Scotland. Dr. Schleifer's experience includes serving as foreman, superintendent, project manager, and partner in a construction company with his brother. He was the founder and president of the largest international consultancy firm serving the contract surety industry in which he assisted in the resolution or salvage of hundreds of distressed or failed construction firms and projects.

This hands-on experience as a contractor and assisting financially distressed companies provided Dr. Schleifer with a unique perspective on the causes of business failure and how to avoid them. His books have been acclaimed by tens of thousands of contractors and are texts in advanced university courses across the country. His weekly newsletter has over 20,000 readers. Dr. Schleifer, often referred to as a "turn around" expert because of the number of companies that he has rescued from financial distress, advises contractors on organizational structure and strategic planning while he also writes, lectures, and teaches.

The importance of education in the construction industry is one of Dr. Schleifer's favorite themes. He has lectured extensively at universities and professional and trade associations, and authored numerous articles on construction business management. Dr. Schleifer has been listed in Who's Who in Finance and Industry, Who's Who in America, and Who's Who in the World, and he was the first Eminent Scholar at Arizona State University construction school.

Books by Dr. Schleifer include: *The Business of Construction Contracting*, Wiley Publishing, 2025; *The Secrets to Construction Business Success*, Routledge Publishing, 2022; *Managing the Profitable Construction Business*, RS Means-Wiley, 2014; *Construction Contractors' Survival Guide*, John Wiley & Sons, 1990. Weekly Newsletter: Schleifer's Weekly Construction Message.

Introduction

The US construction industry comprises of the best builders in the world. Our nation's contractors produce the built environment with ever-improving safety, schedules, and quality. The industry's high quality is a result of the world-class US technical, construction, and engineering education available to it. However, there is limited room in this exceptional technical and engineering education curriculum for finance, accounting, economics, marketing, or business strategy courses. Subjects which are critical to every type of business in the world, including construction. This results in the best builders in the world managing their multi-million-dollar enterprises with very limited *formal business* education. Most are using *instinct* rather than business science. This has led to contractors using "legacy" business processes and beliefs developed over decades and passed down for generations. This is surely one of the primary reasons US construction businesses have the second-highest business-failure rate of industries in the nation.

This book is not a criticism of the construction industry. On the contrary, the author could not be more proud of the industry he has spent a lifetime participating in, and an industry that has produced the built environment from the beginning of time. This book is a result of decades of industry research conducted during the writing of several other construction business texts. During this time, I have collected a list of industry processes, procedures, and beliefs that did not appear to be particularly effective in advancing the profitability of the work. In tracing the origin of these many beliefs, most seem to have been developed through trial and error. I refer to these as "legacy" beliefs because most were effective back when they were first implemented to solve one problem or another. However, there is no evidence that any of these *ideas* were ever *scientifically* tested or verified. Those old beliefs that are still present but no longer produce positive results are presented in these chapters in no particular order.

Drawing on decades of experience turning around failing firms, I define what it will take to advance industry's business practices into world-class performance. The construction industry is slow to change because new entrants are taught traditional practices, which are justified by; "that's the way we have always done it". Few industry professionals recognize the need for change because outdated business practices have simply never been challenged until now. Solutions have never been considered because the unrecognized exposures are accepted as *business as usual.* People don't look for solutions to problems they do not know exist. Construction professionals are not necessarily doing things wrong. They are just not conducting business in the most efficient and therefore most profitable manner. This can obviously be critical in a marginally profitable industry like construction—with the second-highest industry failure rate.

The chapters on project selection, procurement, low-bid mentality, and market cycles provide an enlightened assessment of the many unrecognized or misunderstood risks that contractors face every day. This book includes methods, processes, and procedures to manage around each of the numerous hidden risks. Or, at the very least, describes how to minimize or mitigate the risk

DOI: 10.1201/9781003500315-1

potential. This book addresses complex record-keeping processes and the use and misuse of generally accepted accounting principles that plague the industry. It is my position that there is only one thing worse than a project losing money. And that is a project losing money that you do not know is losing money.

This book tackles head-on the continuing complaints of contractors, banks, and bonding companies alike who are frustrated by positive and profitable job reports that deteriorate over time, and often turn into losses by the end of the project. Ways to determine profit or loss accurately through the life of the projects are described to replace the more typical work-in-process reports that vary up and down each month. I describe how to manage the process of evaluating w*ork-in-progress* in order to correct common misuse, even abuse, of thin critical accounting process. To accomplish this, those responsible need to resist self-serving optimism, and the rewards received for reporting a greater percentage complete. These need to be replaced by *actual* percent complete.

The industry is widely driven by *growth-at-any-cost* despite the reality that profit always trumps growth. This text demonstrates that a major redirection on the topic of growth is a prerequisite to a long overdue need to reduce the industry failure rate. The book includes research statistics, never before presented, that demonstrate industry profits deteriorate as the rate of growth increases from the safe haven of 15% per year. With the inclusion of chapters on equipment costs, joint ventures, market cycles, boards of directors/advisors, and much more, this book is an encyclopedia of profit enhancement and the management of risk.

To manage risk and improve profits construction professionals need to embrace proven business models that work. This book challenges numerous outdated and unsubstantiated industry realities that need to be corrected if the industry is to prosper. Such as:

- Construction industry business competence is weak
- The low bid system makes no sense
- Project selection is too often driven by ego
- *Work-in-progress* accounting is a joke
- Generally Accepted Accounting Principles do not work in construction accounting
- The accuracy of industry accounts payable and receivable is questionable
- Current and prior financial statements should be active management tools
- Project cost control should be separate from general ledger
- There are real limits to growth and growth always increases risk
- Growth is not the primary objective because profit trumps volume
- Management succession is always treacherous
- Joint ventures are always unsafe
- Equipment costs and replacement are always more than you think
- Fully understanding financial and operational capacity is critical
- The dangers of ignoring known and hidden risks
- Recognizing and preparing for industry market cycles is imperative
- The need for non-employee boards of directors/advisors is crucial

Some Things Need to Change

All construction professionals seem to know that there are huge risks in the construction business, and most seem to accept it as a given for which there is nothing anyone can do about it. What has now changed is there is something that can be done about it. Unfortunately, convincing industry participants of this isn't easy because change is always resisted. This book is a frank discussion about how many of the *legacy* processes, procedures, and beliefs the construction industry holds about itself are no longer true and need to be reconsidered. The reasons are simple. The world and our workforce have changed and construction industry management practices, which seem to be random and inconsistent, demand improvement. This becomes obvious when you consider that industry profits have steadily deteriorated over the last 50 years with no improvement in the industry failure rate.

Construction companies are owned and directed by builders and engineers highly skilled at producing the work. However, the majority of these highly skilled production experts were educated as builders, technicians, or engineers with limited exposure to management, business, accounting, economics, finance, human resources, or marketing courses. Most lack any advanced business education in organization skills or business management. It often seems like the preferred construction business organizational structure is "make-it-up-as-you-go". This is a serious industry problem that was masked for decades because, historically, construction was a highly profitable business. Big profits were able to mask shortcomings in the sciences of economics, finance, and accounting. For reasons mentioned later in this book construction profits have been steadily reducing for some time now and show every indication of remaining structurally reduced.

The purpose of this book is to redirect the focus of construction professionals in the direction of business management and leadership. The industry has every opportunity to maximize profits and minimize risk. Reduced profit margins have elevated the business side of the construction business as critical to a firm's success. The construction industry is approaching a crossroads and construction professionals are looking for answers and solutions. This book provides the answers by providing solutions for upgrading processes to deal with project selection, the low-bid process, risk identification, management, and market cycles. These *business issues* (as opposed to technical issues) have become critical subjects for construction professionals in this evolving market. This book also addresses the important subjects of industry consolidation, growth limits, joint ventures, equipment management, succession planning, and boards of directors/advisors. More importantly, a huge portion of the book is devoted to financial and accounting information that most engineers and technicians have never been exposed to.

Whether a general contractor, specialty contractor, construction manager, credit grantor, or student, this book provides a thorough explanation of the intricacies and complexities needed to succeed in the business of construction. As an example, the material includes the controversial subject of waiting for the accounting department to provide cost information on work in progress. Decades ago, the industry decided that project cost control should be supplied by accounting records with cost reports presented to two decimal places covering every activity on the project, no matter how small. The unforeseen consequences of this decision were late information waiting for the monthly closing of the books. This change also put an end to the weekly field practice of "back-of-an-envelope," rounded-off profit calculations for the major work in progress. Some will remember that these *old processes* were timely, and accurate to the degree necessary to manage the work.

At the risk of industry outrage, I recommend returning to this tried-and-true method because it works. Our field personnel desperately need timely cost information, and they do not need to be accurate to two decimal places or available weeks after the work has been put in place. This is an area that demands going back to what worked in the past because the accounting department rightfully does not care what the costs are, as long as they are totally accurate. While the field people do care what the costs are, but only care if they are timely enough that the field can do something about them. To manage risk and improve profits, we need to embrace proven business models that work.

Key Feature—Overview

There are two distinct sides to the construction process: the performance of the work and the business side of the business of construction. This book is about the business side of construction contracting and defines what is currently working, what is not working, and what needs to be done about that. It provides processes, procedures and practices that will advance any construction enterprise towards the future. The construction industry is approaching a crossroads. Construction professionals have questions and are becoming desperate for answers and solutions. Drawing on decades of experience turning around failing construction enterprises, *legacy* industry practices are identified. Many of these are obsolete to the degree they are actually magnifying risk.

This book will:

- Redirect construction professionals from concentrating solely on building things to a greater understanding of the *big picture* and how the industry, and, therefore their company, actually functions.
- Provide detailed solutions to outdated processes and procedures that will immediately enhance profits and minimize risk.
- Demonstrate how some of the time and energy spent on producing the work must be redirected to an increase in the time and energy put into producing a profit.
- Offer access to numerous templates and tools developed for the author's consulting practice.
- Provide text and reading for graduate or undergraduate courses in management and strategy, accounting, financial analysis, and marketing for construction.

Construction has a limited number of professional books devoted to it compared to many other industries. And almost all of these books concentrate on the field side of construction, regardless of what their title implies. In my experience construction professionals, particularly top management, focus the majority of their time and effort on getting the work done with far less time being spent on the business side of the business. Some have said: *If you take care of the field, the business will take care of itself.* **That is simply not so.** A lack of attention to, and competence in, the business side of the business of construction is the primary reason that the construction industry has the terrible failure rate. The failure rate is forcing construction professionals to recognize that the industry is evolving and construction risks are increasing while profits are decreasing.

Contractors recognize that family members are not joining family construction businesses, and field people are not encouraging their children to join the trades. The industry has become clearly less attractive, and participants are asking why.

Decades of trying new contracting procurement methods like Construction Management, CM at Risk, Design/Build, Progressive Design build, Partnering, Public-Private Partnering, Best Value, Integrated Project Delivery, and numerous variations of these have not satisfied owners, designers, or contractors. Nor have any of them consistently reduced losses or disputes. Some owners are even reverting back to low bid, which is what started the quest for a better procurement method in the first place. Industry professionals are asking *what's next?*

Contract documents are becoming more lopsided in favor of the owners than ever before, which simply fosters disputes and litigation. This is what caused the *procurement revolution* to begin with. All of the parties concerned are asking why aren't the new procurement methods working and what's next? What's next looks like increased complexity of construction regulations, more complex OSHA requirement, challenging contract language, increasing labor shortages, and spiraling insurance costs with limited availability. These issues are driving overhead up at a pace that defies recovery. This compounds the problem that next year's overhead expenses are never fully included in this year's estimates. No one seems to recognize the inequity of contractors bearing all of the risk of inflation that cannot be known beyond a guess. Some are asking; *If we can't know the unknown why should contractors have to pay for the unknown?*

This book offers the solution to this and the other outdated industry practices that need to be changed.

1 The Low Bid System Makes No Sense

Introduction

Decades ago, the federal government legislated a lump sum bidding system for federal procurement, including construction, that rewarded the lowest price submitted. The process is more or less an auction. After signing the contract, what it actually costs the contractor to build the project became meaningless and of little or no interest to the project owner or designer. The construction industry quickly achieved the second highest failure rate of any industry in the nation.

Some History

For decades after the founding of the United States, the procurement processes for material, supplies, and services drew heavy criticism. Finally, in 1809 congress enacted legislation requiring all contracts for supplies and services, including construction, be awarded to the lowest bidder. That legislation was expanded on in 1909 when the federal government produced a standard form of contracts and bonds (Defense Contract Management Agency).

In 1994 congress passed the Federal Acquisition Streamlining Act (FASA) which described a "preference" for the acquisition of commercial items in meeting agencies' needs. FASA also articulates composition requirements for task order and delivery order contracts (Congressional Research Service). Prior to the 1990s Public (Federal, State and Municipal) construction contracts in the United States were 99.9% low bid contracts. Even the rare cost-plus contracts were awarded based on competition for the fee amount, either a lump sum fee or a percentage of the costs incurred.

The low bid process clearly placed the contractor at "arm's length" from the buyer of construction services and the designer of the project. The public bid-opening procedure allowed the "winning" contractor to see that their price was lower than all of their competitors and by how much. An independent observer might consider the winning contractor in an almost *frightening* position depending on how low they were compared with their competitors' proposed amounts. Obviously, the low bidder would have a legitimate and logical concern about protecting whatever profit they anticipated in their low price (Figure 1.1). The level of concern would certainly be impacted by how much lower the winning bid was from the next lowest bidder or from the average of several of the lower but unsuccessful bids.

There is some science in quantifying the work and estimating the costs but also some guess work, particularly in how much overhead to apply and how much markup for profit to include. *As a contractor, I have been at the reading of bids in which I had estimated, priced, and proposed on the project in question. And when several bids higher than mine were announced I would wish I had increased the amount of my as yet unannounced bid. However, when a bid was read that was lower*

DOI: 10.1201/9781003500315-2

Figure 1.1 Public bid openings emphasize price competition but leave contractors vulnerable to risk and uncertainty.

than mine I would all of a sudden be wishing I had bid less. What I came to realize over time was that no matter how hard I worked at estimating and how confident I was about pricing the work, when all the other prices were announced the anticipated cost of the work and the value I placed on it was a matter of opinion.

So herein lies a part of the problem that many outside the industry, and too many inside it, fail to understand that **cost and price are not the same thing**. Cost is an internal approximation while price is a strategic position. Estimating the cost is a discipline while setting the price is a business decision.

With the immense amount of detail that goes into preparing an estimate, even on smaller projects, the potential for mistakes is enormous. Mistakes happen and even I am embarrassed to admit that some of the projects that I *won* as low bidder were a result of my mistakes, not my superior estimating skill. *(Any readers that can relate to this will understand why I almost titled this chapter "The Low Bid System Is a Joke".)*

Some contractors will claim total confidence in their pricing and bid numbers, but I can assure readers that they are the exception. Webster's Dictionary defines an *estimate* as *a rough calculation of size, value, etc.; an approximate computation of the probable cost of a piece of work; or an opinion or judgment.* Professional estimators and contractors often describe their estimates as *a calculation* or a *computation.* I fully understand the reasoning but respectfully suggest that this level of confidence is an error that our industry needs to correct. Webster uses the same words with qualifiers as in *a **rough** calculation* and *an **approximate** computation.* To describe an estimate, even computer generated, as a "calculation" or a "computation" is not just wrong, it is dangerous.

In defining *estimate*, Webster's uses the words *rough* and *approximation*, which are closer to a guess than a fact. If a reader gets nothing else from this book, please understand that an estimate is not a *calculation*, it is a ***rough calculation***; and that an estimate is not a *computation*, it is an ***approximate computation***. (While the estimated ***cost*** of the project may be the basis for the bid price, it may not remain the ***price*** a contractor may end up agreeing to do the work for.)

Questioning Low Bid

During the 1990s there appears to have been a realization among some buyers and designers of construction projects that low bid contracts were not providing the buyer with "best value". Schedule was also a factor because project design had to be 100% completed before contractors could calculate the total costs for the project. And for projects of any size, it usually takes as much as a year or more to complete the design before construction can begin. Speed to market became a greater concern as, nationwide, the pace of commercial transactions was accelerating. Delivery of the built environment needed to be fast-tracked. This in turn accelerated the development and introduction of new contracting methods such as:

- Fee-Based Construction Management
- Construction Manager at Risk (CMAR)
- Guaranteed Maximum Price
- Design-Build
- Integrated Project Delivery
- Job Order Contracting
- Private-Public Partnerships

Many outside the construction industry do not seem to understand that construction contractors don't participate in the selection of the contracting method on their projects. The contracting method is chosen by the owner of the project who normally seeks input from the designer they have selected. This may seem appropriate in that just about all construction activity begins as a need on the part of the end user of the project (the owner), such as the need for a building, a road, a sewage treatment plant, and such. The owner with such a need then selects a designer qualified in the type of project. In the traditional low bid contracting method, used almost exclusively prior to 1990, after the design was completed, the project would then be put out to bid.

Beginning in the 1990s, the construction industry evolved to the extent that the buyers and designers of construction projects began to dictate different and varied delivery methods. The above-listed contracting methods were introduced as an alternative to the conventional low bid method attempting to provide the project owner with greater value. Most of the alternative methods require that a contractor be brought into the process during the design phase. Contractor selection during the design phase necessitated the selection of the contractor on factors other than the price of the as yet designed finish project. The new contracting methods did not immediately satisfy the very different needs of the owners and designers who were dissatisfied with the low bid process.

The new methods did not necessarily benefit the contractors either. Consequently, multiple variations of the new contracting methods were tried, and numerous permutations of the various methods were experimented with. After two decades of experimenting with the contracting process, which some may look back on as "confusion" within the industry, no one contracting method surfaced as satisfying all participants. And then, during the industry slowdown (recession) of 2009–2010, surprising most industry professionals, a number of municipal and state public bodies began to revert to low bid contracting methods attempting to maximize what they could get with their limited budgets. Some private buyers followed suit, but the attempt to switch back to low bid never really caught on and there was much speculation as to why low bid was being attempted again.

In the author's opinion, the contracting-method changes were a result of many contracting officers and owners' representatives' feeling that they were losing control of the construction process and that, under alternative-delivery contracting methods, contractors had undue influence over project decisions. Some owners' representatives and contracting officers felt that there was not enough transparency in decision making and design changes during both the design and construction phases. (Some still do.) The consensus seemed to be that bringing contractors into the project earlier would result in more of a "best value" contracting method.

Questioning New Methods

To date, the many new and varied contracting methods have not satisfied all of the wants, needs, or desires of the parties involved. Contractors did not fair very well during these decades. Research indicates that the average profits across the country did not improve and in some areas reduced. Most telling of all is the failure rate among construction contractors did not improve at all and the industry remained the second highest failure rate in the nation, second only to restaurants. A Travelers Bonding Company study indicates that over 50% of heavy civil projects suffered profit fade and went on to say that the construction industry has not been able to price risk. A PricewaterhouseCoopers study states that only 25% of (construction) projects are defined as successful in scope, cost, schedule, and business. In 2015 a Construction Industry Institute (CII) Performance Assessment Committee reported that only 30% of (construction) projects completed within 10% of the planned cost and schedule. And in 2025, study reported a 24% growth in owners' construction costs since 1995—net of inflation and escalation. Clearly, decades of experimenting with construction contracting methods have produced little for contractors. There was some advantage for buyers of construction services in that overall time for design and construction has improved.

The real test of the new methods is earnings. The majority of owners, designers, and contractors are for-profit organizations and for the most part corporations. (Clearly, there are some not-for-profit buyers of construction services but other than government bodies, they represent a small percentage.) For any contracting method to work efficiently, the designers and producers of the product must earn a fair profit and the owner of the project must receive fair value. For any commercial process to function well, it must be equitable to all parties. It makes little sense, and to my way of thinking, it is not sustainable, for one of the largest and most critical industries in the United States to have the second highest failure rate. It comes close to a national security issue that the *structure* of the construction industry places an illogical and unconscionable amount of risk on the contractor that produces the project (Figure 1.2). Some risk, yes. But all the risk, no. The contractor should benefit from a reasonable profit and perhaps the risk of a reduction in profit for weak performance. But the low bid system not only risks loss but obviously and undeniably risks business failure. *Something is wrong.*

Contract Bias

One of the reasons that low bid contracting method did not work well is that by its very nature it created an adversarial relationship. Attorneys for the owners or designers create contracts that govern payment and performance. The attorneys apparently believe their assignment is to protect the owner (and the owner's agent—the designer) from any and all project risk. To do that, the risk has to be placed somewhere. There is only one place to put it—on the other party to the contract, the contractor. Starting in the latter half of the last century, the

Figure 1.2 Alternative delivery methods involve contractors earlier in the design process, but unproportionally shift project risk to the contractors.

contract documents between the owners and contractors became more and more protective of the owners. This was unfortunately at the expense of the contractors. Prior to WWII, standard construction contracts were reasonably equitable and fair to all parties. In the 1970s the revisions to standard construction contract documents began to slant in favor of the owners and designers. That trend continues today, with an increasingly strong and irrefutable bias toward the owner.

Some History

At the risk of dating myself, I wrote an article for the January 1974 Constructor Magazine titled Responsibility without Authority which mentioned what was just the beginning of bias in the standard American Institute of Architects (AIA) construction contract documents. The article mentioned the first known set of written laws governing, among other things, construction. More than 4000 years old, the Babylonian Code of Hammurabi was found written on stone tablets. The code stated: *If a builder constructed a house with the result that his work collapsed and so caused the death of the owner of the house, that builder shall be put to death.*

In that 1974 article, I said: Today the lines of responsibility in building contracts are often unclear, muddied by ambiguous language. You would think things would have gotten better, but unfortunately, today that statement is more than accurate and still applies. The article stated that the AIA contract language placed most of the authority over the construction process in the hands of the designer while placing most of the responsibility for the construction process with the contractor. This is still true today and even more so. This irrational and unreasonably imbalanced placement of risk is a big part of the reason the construction industry has the second highest failure rate in the nation. The reason this cause and effect has gone unnoticed is that when a project goes reasonably well, which is the case with most construction projects, these irrational and unreasonable clauses in the contract are mute and do not come into play. However, when anything goes wrong during construction, no matter the reason, the contractor is contractually responsible.

Non-Negotiable

What makes the bias in construction contracts even worse is that the contracts are non-negotiable. For readers not directly involved in the construction process, some background will be helpful. Construction contracts for all lump sum low bid projects, and most other contracting methods, are prepared in advance of any construction activity by attorneys for the owner and/or designer. The contract documents are included in the bid documents, and they are identified as non-negotiable. That means that when a contractor accepts the plans and specifications for estimating and then submits a bid, the contractor is in effect accepting the contract documents as distributed by the owner. Bid documents state that any proposal submitted that is not in conformance with the bid documents, which includes the entire contract, is automatically disqualified. This obviously declares the contract documents non-negotiable, so any attempt to object or modify a contract clause disqualifies the bid. The contract is non-negotiable, period.

Construction contracts being non-negotiable may very well have led to the accelerated and aggressive manner in which the construction industry standard forms of agreements have been continuously re-written in favor of the owners and designers. There are three parties to the construction process—owner of the project, designer, and contractor. The owners and designers benefit from the unbalanced risk distribution and therefore cannot be expected to object to the risk allocation. Contractors are, and I hesitate to use the word, the *victims* of this deliberate and accelerating unbalanced risk distribution. Contractors and their trade associations are the only ones who have any motivation to fight to change this injustice. Trade associations do a great job of advancing their members' collective needs, which sometimes conflict with other trade associations' positions. Trade associations are unfortunately not known for cooperative efforts with other trade associations. However, a collective assertion of the industry-wide benefits to all parties to the construction process from an equitable distribution of inherent risks could be a rallying cry.

Mutual Interests

Cooperation among trade associations is impacted by some outdated beliefs that need to change for the good of the industry. Some trade association professionals have said that their trade members are highly specialized and have very little in common with other trades. For example, I have often heard it said that electrical contractors have nothing in common with plumbing contractors, or that heavy highway contractors have nothing in common with building contractors, and so on. There is a reality in the modern construction industry that does not seem to be commonly understood. Plumbers, painters, and plasterers may be very unique and different trades; each having specific technical issues of interest to their type of work. However, they are all operating businesses that provide services within the same industry—construction.

It will prove helpful for construction professionals of every trade and every size of company to see and understand their enterprise from an analytic perspective. If you look at your company's field activity, you see clearly the technical side of the business in action. However, when you look at the office side of the business, you only see desks and computers. Historically, construction people consider technical issues and production to be *the entire business* and typically consider the office a support activity. Back when industry profits were double digits, an organization could get by with that belief. With the reduced profit margins of the modern construction industry, a more analytic understanding of the business functions is critical. I use the term *critical* deliberately because my experience is that none of the construction firms that have absorbed these realities and put this knowledge to work in their organizations have failed. This is significant in an industry with the second highest failure rate in the nation.

Business Functions

To be successful in the modern construction industry, you must fully understand that there are two very distinct sides to a construction operation. They are the technical sides and the business sides of the construction business. They are totally distinct and stand alone. They complement each other and are equally critical. Many construction professionals do not understand this, so it bears repeating. The technical sides and business sides of a construction business are equal in importance to the success of the organization. If an organization is poor on the technical side, it will fail—period, end of discussion. It will fail likely sooner than later, but it will fail. The failure rate of start-up construction companies is off the charts based on my 50 years of research on failed construction firms. Companies under ten years of age were never included in the data. The reason is the failure rate was so great; it would skew the data of established companies.

If established construction firms, meaning they have been in business for over ten years, remain competent with the technical side of the business, they will fail only if they make mistakes on the business side of the business. For this reason, some consider the business side more important than the technical side. I measure each as equally important because if a group is poor at the technical side, they were never really a *business*, no matter how long they lasted. To be successful as a construction contractor in any trade, or as a general contractor, the organization must be capable at the trade or trades they perform, which of course is the technical side of the business. If the organization is not at least minimally proficient, they are by definition not a *contractor*. They may attempt to function as one, but they eventually fail. Therefore, the technical side is a *prerequisite* for success, and being capable at the business side may seem secondary. The reality is that being competent at both the technical and business sides of the business is critical for success.

Contractor Training

A dilemma within the construction industry is that it takes so long to become proficient at the technical side of a trade or general contractor business, which is accomplished through either:

- apprenticeship and work experience
- or technical trade school and work experience

Apparently, there is never enough time to also include a business school education.

What is also true (*a catch-22 you might say*) is that an individual highly skill at business could not easily become a construction contractor in a small or midsize operation. This is true because a contractor, even one who has no need to actually work with the tools, cannot possibly succeed if they don't know the trade they are selling. They could not supervise production and would not be qualified to supervise the people they may hire to supervise production. It will be helpful to be reminded of the definition of *small* and *midsize* construction company.

A small construction company is one in which the contractor directly supervises all the projects and may or may not work with the tools on occasion.

A midsize construction company is one in which the contractor manages those who directly supervise all the projects and periodically visit the project sites.

The point here is that for the small and midsize construction firms, the production of the work at a profit is a primary requirement for success. In these two sizes of construction organizations, small and midsize, the contractor (boss) or one or several partners with ownership (bosses) are personally responsible for the generation of profit—the lifeblood of a business. To better understand this, let's contrast this with a large construction firm.

A large construction company is one in which the contractor (or partners) spends 90% to 100% of their time managing the business and has no direct supervisory responsibility for any projects. If for any reason they visit a project site, it would not be to supervise any work.

Clearly, the size of a construction firm matters in terms of organizational structure and responsibilities.

Management Structure

In small and midsize construction firms, the contractor supervises projects or supervises those that supervise projects. In the small firm, there is limited time for the contractor to manage the business side of the business, but fortunately, the business side of the small construction firm requires less time. There is no clear-cut definition but on average I find it to be about 75% to 80% production effort and time and 20% to 25% business effort for the small firm. This will vary from month to month, for example, during tax season or when extra marketing seems appropriate, more business side time may be required. The important thing is to recognize that a contractor must in effect wear two hats—technical or production side and business side. It is more important to understand that no one can wear two hats at the same time.

It is difficult for many to understand that the production and business sides do not overlap. It is not necessary to set aside specific hours or days to address either side, but it is also not possible to switch back and forth between technical and business issues minute to minute because in doing that, nothing gets done well. When this reality is first introduced, almost everyone argues that there are constant overlaps and separating the sides is impractical or impossible. I am bombarded with examples like budgeting a project is a business issue. It is not. Just because it is a numbers or accounting function does not mean it is a business side issue. It is purely a field side issue. I get the rebuttal: suppose there is not enough money in the estimate? It remains a production issue. The estimate is past tense. You can't go back and re-bid the estimate. Then I hear, but it's not my fault. Managing is not "placing blame" which is best described as a waste of time.

When functioning on the production side consider production issues exclusively and likewise when functioning on the business side consider business issues exclusively. It takes some time to fully understand that there is no overlap because most construction professionals believe that the production and business sides function together. They also combine information from each other in attempting to solve problems which often leads to erroneous solutions. An example: While working on the business side reviewing overhead cost reports, waste is discovered on a particular line item. The inclination is to bring shop or field people into the review and attempt to develop solutions. The waste discovered is past tense. Sending a notice to production people is an appropriate business side activity, but solving the problem is exclusively a production side activity. Directly affecting change is production side activity that can be done now or in the future. The person discovering the problem may or may not be involved in the solution and if they are, they will then be wearing their production hat. There is no *overlap* and continually changing hats is ineffective and an interruption. Affecting change (solution) can just as easily be done later.

Another example: An owner's representative approaches the project field superintendent expressing an idea about relocating a doorway in the next phase of the work. The superintendent stops what they are doing, goes over to the field trailer, and gets out the prints to gather the details and discuss the pros and cons and agrees to set up a meeting with the owner's representative, the project manager, and him or herself. A meeting subsequently takes place, and the superintendent is permanently involved in this potential change that may or may not be decided upon. This happens all the time and many readers will be saying it is the correct and obvious way to deal with these types of issue. Herein lies the problem. It seems logical and is the way these types of issues are handled in our industry. *This may be a good spot to mention that just because it is commonly heard in construction, "This is the way we always do it", does not make it is right.*

The superintendent is always busy and key to project progress and profitability. His job description includes interaction with the owner's representatives and a good superintendent who establishes an excellent rapport with them. A good superintendent also understands his job description, is 100% focused on producing the work at a profit, and is not easily distracted. When asked a question that is outside his job description, like a request for a possible change in the work, he or she professionally says *I will have the project manager get in touch with you as quickly as possible concerning that*. The modern construction enterprise has well-trained employees who know what they are doing and are not easily distracted. (The first time this occurs the superintendent might have to add, *I have nothing to do with changes in the work.*)

A construction organization's efficiency and productivity is its own responsibility and the modern construction enterprise is organized accordingly. Any inspector or owner's representative will tell you that their influence is limited by the level of the superintendent's knowledge, professionalism, and confidence. And inspectors have been known to admit that they take advantage of weak or uninformed superintendents.

Summary

The industry is evolving, profit margins have deteriorated as far as they can go, and construction is close to becoming a commodity. Construction professionals consider our services a custom service that varies from contractor to contractor based on their qualification for any particular project and influenced by the project's size and type. Most construction professionals also believe that a contractor's qualifications for a project can be measured from least to most. However, today's buyers of construction services seem to believe that construction is a commodity and that for the most part, one contractor is the same as another. That was certainly true during the low bid days when contractor eligibility was screened by the size of their balance sheet. Along with the development of the many different types of contracting methods mentioned in this chapter, various and multiple screening processes have been and continue to be created. Considering that there has been no improvement in the construction business failure rate, the effectiveness of the new screening processes may be in question.

While contractors may not be able to alter the ever-changing project selection systems and processes, they need to alter their behavior to avoid the serious consequences of taking cheap work. The major driver that needs to be controlled is to resist the self-imposed mandate to maintain size or increase sales. The drive for growth at any cost is suicide, especially if it means taking on unprofitable work. The appropriate (and perhaps only) defense is controlled and skillful project selection, which is explained in detail, including access to free software to accomplish it in Chapter 2. Case studies will demonstrate graphically the dramatic risk and consequences of business failures and bankruptcies.

2 Project Selection Is Too Often Driven by Ego

Introduction

There are no bad projects, just bad matches of contractors to projects. The risk associated with project selection can be accurately measured in advance and is directly associated with the construction organization experience with similar work. Therefore project risk differs for each contractor. The performance of an operating entity, such as manufacturing, improves with repetition. In contrast, construction enterprises do not usually have enough repetition from project to project to experience much improvement because in almost all cases, new projects differ in varying and measurable degrees from previous projects. The closer the new project represents the average of previous projects, the more likely estimated performance will be achieved, simply because the planning and execution is more of the same as opposed to totally unique. The measurement of risk also depends on the experience of the team members involved. Experience is accumulated institutionally but is captured individually, so the number of members on a performance team with direct experience on similar previous projects impacts the likelihood of estimated or improved performance.

When the Market Chooses You

It almost seems out of place to use the term *project selection* within the construction industry because most contractors I have encountered don't relate to it. When asked how they select the projects they claim to pursue the work that is available and they don't actually "make a selection". Using this process effectively means the market selects them. When asked what specific size and type of work they go after, most react curiously to the question, as if they have never given it any thought or that the answer should be obvious. I hear responses like, *I'm an electrician so I go after electrical work, I'm a plumber so I go after plumbing work*, or *I'm a highway contractor so I go after highway work*. These answers imply that there is only one kind of electrical, plumbing, or highway work. Exploring further, I'll usually ask: *What if after you obtain the plans, you find there is something unique in the plans or specifications that you have never done before*. The response is usually along the lines of *that's no problem, I'll figure it out*.

There seems to exist an assumption among many contractors (I genuinely hope not most) that after being in business for a while they know all there is to know about their trade. And while a huge number of contractors say they regularly run into things they have never seen before, most describe those occasions *as a great learning opportunity to advance their business*, or similar type of reasoning. Some have actually said that these learn-as-you-go opportunities are great because they happen at the expense of the owners of the projects, and they (contractors) consider it a bargain. These contractors seem to ignore the reality that if they price work that they have never done before too low, which is something that we all know happens, it will cost them, not the owner.

DOI: 10.1201/9781003500315-3

While all contractors are supposed to be experts in their trade or trades, it concerns me that this *learning* includes discovering what it costs to perform work they have never done before. There is obviously huge risk in estimating the cost of work you have never done before because, without experience, it is difficult to be confident you know what the cost is going to be. Obviously, you will learn the actual cost of the work when you perform it, but of course, by then it is too late to change your estimate. When I ask, *isn't pricing work you have never done before really a guess?* The answer is usually some form of *How bad could it be?* or *How else am I going to get the work?* or *This is the only way I am going to learn something new.*

A Growth-Driven Industry

There is an art and a science to Project Selection but convincing contractors to believe that, let alone use the science and develop the art, is a hard sell. Having discovered the limited interest contractors seem to have in the topic, I tried to ascertain why. After years of inquiries the only reason I have discovered is that contractors see the process of project selection as restricting their company's growth in sales, and sales seem to be all that contractors care about. Continuous growth in sales appears to be almost a universal *prime directive* for contractor/entrepreneurs. Contractors that I talk with seem to believe bigger is better with many saying in one form or another, *success in business means being the biggest*, with some saying, *biggest and the best*. After hearing this so many times, I continued to research the issue. Over the years, I became convinced that, consciously or unconsciously, this is a prime driver of most small and mid-size contractors and certainly for most large construction firms. Growth seems to be a driver despite the reality that small construction companies make a greater percentage of profit than mid-size organizations, who in turn make a greater percentage of profit than large construction companies. Because a lesser percentage of profit implies at the very least a greater risk, you would think that construction professionals would see growth as having pluses and minuses, or even as being high risk and approach growth more cautiously.

There have been a number of articles over the years and books by this author that provide conclusive evidence that in construction, growth in and of itself includes risk and that rapid growth magnifies that risk. My research indicates that the safe rate of growth in the construction industry is 15% a year and that continuous growth at that rate is sustainable at minimum risk. Data from a statistically significant number of construction firms indicates that companies with a rate of growth of 15% a year or less had no failures. Those growing at 25% a year had some failures and those growing at 50%, 75%, or doubling in size in a year had progressively greater failure rates (Figure 2.1). It is fair to assume that this information was not known by the firms that failed while growing too fast or they would not have failed. It is almost certain that the firms that failed during excessive rates of growth did experience the stresses and strains that rapid growth places on an organization and had to notice diminishing cash flow. These are signs reported by every failed firm investigated.

Ego Is a Strong Driver

The question arises: What are the drivers that cause a rational businessperson to undertake rapid growth and then to continue rapid growth after experiencing the serious stresses and strains that we know occur during extended periods of rapid growth? The initial decision to grow the business may be a belief that it is a smart, even logical thing to do. However, when the stresses and strains present themselves, as we know they do, why then continue the growth? It would seem that

Figure 2.1 Relationship between sales volume growth and business failure.

logic would dictate that something was wrong, and that a reasonable businessperson would slow or discontinue growth to determine the cause. Having asked this question of hundreds of failed contractors and carefully evaluating the responses, it appears in the majority of cases (and possibly all, as some cases were indeterminant) that business reasons did not appear to play a major part in the decisions about growth. It appears that *ego* was the primary driver, which is information that needs to be explored further.

The definition of "egocentric" is:

> *Dwelling upon oneself or upon one's own personal interests almost to the exclusion of everything else; viewing everything in relation to oneself.*

As most people will agree, entrepreneurs (like contractors) are almost universally type A personalities. This suggests that my comment in the previous section of this chapter, *that after being in business for a while, they* (contractors) *believe that they know all there is to know about their trade*, may be correct. Type A personalities are self-confident and those that prefer being their own boss tend to go into their own business. Contractors fit this definition and those that start their own business and succeed are truly *self-made*. Some describe being in their own business as lonely in that there is generally no one in their employ that is capable or that they trust to bounce ideas off. So, if success is achieved, a self-made contractor realizes that they actually achieved it on their own. That is naturally quite satisfying and feeds the ego, which most type A personalities have plenty of. This is not meant to be disparaging of contractors, but rather complementary. What it does describe is a path to trusting one's own judgment and an explanation for why many contractors do not tend to seek outside independent advice. Which can be both a strength and a weakness.

Technical vs Business Strength

Strong self-confidence can become a serious weakness as an organization grows from small to midsize. At midsize, a contractor can no longer run the business alone. They are no longer closely associated with the field in that others responsible for field operations report to them. As such, contractors no longer have firsthand knowledge about field operation. The contractor's decisions are at this point more about the business side of the business and less about operations. For an organization of this size, business decisions become much more important, and in many cases critical, to the success of the firm. The result is that small construction business success is almost totally dependent on the contractor's ability to put the work in place and those that make it to mid-size are those who are most skilled on the technical side of the business. Few, if any of those educated or trained in the trade side of the business that have successfully caused the work to be put in place at a profit, could possibly have had the time for a business education or to have accumulated the experience.

Herein lies the problem. Successfully starting and running a small construction operation is often totally dependent on strong *technical skills*, and in no way prepares a contractor to manage a mid-size construction organization. Successfully running a mid-size construction organization is almost totally dependent on strong *business skills*. The bigger problem is that most construction professionals do not know this; therefore, they have no reason to be concerned or to take any corrective actions to address the problem or to recommend support to the unprepared candidate.

Can Ego Be an Issue?

I fear that some readers may think that this is too strong or too extreme of an assessment, but I would remind readers that construction is one of the largest industries in the nation and that it still maintains the second highest failure rate in the nation. In 1987, I published the common elements of construction business failure and have since then continued to research why the failure rate continues to remain so high. There is nothing wrong with the product that our industry has been providing which produces the built environment, from the pyramids of Egypt to the skyscrapers of today. The technical or production side of the industry is strong and is not the reason for our failure rate.

100% of the causes of the construction business
failure rates occur on the business side of the business.

Many have claimed that size, type, or location of project could not possibly be common elements of construction business failure. However, the research proves conclusively that these elements regularly and continually cause business failure. These are not field issues because no one in the field decides what size, type, or location of projects to go after. These are purely management business decisions. Weaknesses on the business side of the business of construction originate in overconfidence (ego). The good news is that the cure for inappropriate decisions in this area is not complicated. It simply requires recognizing the exposure exists.

Elements of Construction Business Failure

The common elements of construction business failure are as follows:

- Increase in project size
- Changes in type of project
- Changes in geographic area
- Changes in key personnel
- Lack of managerial maturity

All the failed construction companies we've studied experienced one, or in many cases more than one, of these elements. The primary cause of construction business failure is growth, and some might even refer to growth as a *natural* reaction to market conditions. When there is a lot of work around, most would expect construction companies to grow. Conversely, construction companies facing a down market will often expand the size and type of project and location of the work to resist negative growth. That motivation may also be described as *growth* because it is a refusal to get smaller or a determination to remain the same size in a shrinking market. In either case, contractors are likely to engage one or more of the common elements of construction business failure. Contractors who expand their size, type, or location of projects will naturally have to add key personnel, which is another common element of failure. Some may simply be convinced that growth is the smart thing to do, but it is hard not to believe that *ego* is involved, in whole or in part, in decisions to grow. The hundreds, if not thousands, of contractors I have encountered have stated or exhibited a desire, and in some cases an extreme desire, to grow their businesses as large as possible and as quickly as possible.

Many construction professionals have asked me if there are any compelling or indisputable restrictions on the rate of growth? A hard-driving contractor/entrepreneur may feel that there are no practical limitations to growth or may not understand some of the financial parameters of growth such as cash flow impacts and so on.

> *Without some formal business education, a contractor would not likely know of the 15% limit on safe grow in their industry or the very real dangers of rapid growth.*

Motivated by Growth

Questions about growth come down to what is *the motivation for growth*? It is hard to ignore the likelihood that a contractor with a desire to grow their business may unconsciously or consciously be driven by ego. We all have ego, and it is certainly not a negative characteristic unless it is out of control. Ego is also regularly associated with motivation. I submit that without formal business education, the risks associated with growth and the limitations on rate of growth would be unknown.

Contractors could, of course, consult with others about growth and rates of growth and there may or may not be managers within the company that the contractor could ask. However, contractors are not known for being particularly inclusive in decision making. There may also be people associated with the company that could be consulted with, such as the outside accountants, bankers, and so on, but the same reluctance seems to apply. Other contractors or members of the trade association may be another source of information; however, small and mid-size contractors don't usually test their ideas with their competitors. The norm in our industry seems to be that decisions about growth are, for the most part, made in a vacuum and often in a hurry. This leaves only the personality characteristic of ego as the likely motivator for a contractor's growth with few others, if any, to advise on those decisions. This appears to be a high-risk area made even more risky by a lack of checks or balances.

Contractors are busy and make dozens of decisions on any given day. If they have been in business for any length of time, they naturally grow confident in their decision-making ability because trusting their own judgment is how they got where they are. Unfortunately, this may be part of the reason we have so many examples of good contractors finding themselves in financial difficulties. It may also be one of the reasons the construction industry has the second highest failure rate in the nation. Here are just a few construction business failure examples that reinforce this rational. Readers should keep in mind that this data includes only contractors that have been in business for at least ten years:

- A masonry contractor won the bid for the brickwork on a large commercial building in his area. He had never put glazed brick in place before and ran out of money before he could complete the project.
- A successful sewer pipe contractor in business for many years took his first ductile iron force main sewer contract that had to be completed by his surety company after the project bankrupted him.
- A bridge builder in business for over 25 years took a contract for a bridge over a body of water. It was the first bridge the company had built that crossed over water and the last bridge the company ever built.

The examples are endless, and I can assure you that when these firms decided to pursue these projects, they did not think the jobs were any different or any riskier from the work that they had performed in the past. They felt it was their type of work; otherwise, they would not have pursued it. During my 60 years of construction experience, I can only point to one contractor that could describe in great detail the exact type of projects their firm went after. None of the thousands of other firms I interacted with or researched had a definitive description of the work they pursued. Most, if not all, contractors have a vague depiction and deliberately wide variance in the work that they are interested in. Many contractors truly believe *we can build anything*. That statement is concerning because the question that all contractors should be asking themselves when considering a potential project is not *can we build it*? The question is, and always should be, *can we build it for a profit?*

What follows is a more in-depth discussion of some of these common causes of construction business failure. All these causes are more completely described in *Construction Contractors' Survival Guide*, John Wiley & Sons, 1990, and also in *The Business of Construction Contracting*, John Wiley & Sons, 2025.

Increase in Project Size

When I ask contractors about what size project they specialize in, the most common answer is a response like *the bigger the better*. This always concerns me because my decades of research on the cause of construction business failure proves that a change in size of projects is the primary cause of construction business failure. Growth and rate of growth are critical factors in business failure, and yet rapid growth is a primary driver for most contractors.

While earning a profit should be the primary business objective, the antithesis of *profit-driven growth* can be used to describe the huge number of contractors who allow ego to be the driving force behind their business philosophy with growth as the ultimate goal. A business philosophy or ultimate goal of growth can't work because profitable growth in this industry has practical limits. Continuous growth is particularly dangerous because even unprofitable growth generates cash flow, and positive cash flow can mask the lack of profit for a long time because this month's receipts pay last month's bills. The reason is that each month during a growth phase, the organization invoices its clients for more money than the expenses they incurred the prior month. This continuously increases the contractor's cash flow but does not necessarily increase profits. A dangerous problem arises because if growth slows or stops for any reason, the cash flow dries up rapidly and any weakness in profitability is immediately exposed by the inability to pay bills. If there is no alternative source of funds, insolvency may force bankruptcy.

The risks inherent in continuous growth can be mitigated by a firm returning to slow or no growth after two or three years of growth. This allows for a serious and detailed evaluation of performance and profitability. Accurate evaluation is impossible during continuous growth because during growth periods, everyone is too busy for a serious evaluation. Preparing and evaluating critical numbers, like the cost to complete a project, take significant time and no one seems to have time available. Accurately tracking and measuring profitability during continuous growth is almost impossible because positive cash flow masks profit shortcomings and there are just too many new employees and too much activity to track or measure anything. Anyone who has experienced rapid growth in a construction operation, for example, doubling sales within a year or two, can describe the *frenzied* pace of such business operations. Controlling the drive for growth is much easier when one learns how dangerous continuous or rapid growth can be. How to accurately analyze performance in a growth environment is presented in Chapter 10, including case studies, financial parameters, and formulas.

Moving into larger projects regularly occurs during profitable years and problems can develop even before a large project is completed. Profitability and potential for loss are directly related to the size of a project compared to an organization's typical projects. Being profitable at current size projects is not an indicator that a firm will automatically profit attempting larger work. The question is not, can you build it? Rather can you build it at a *profit*? As project size increases, so does the strain on the company's resources, technical abilities, and collective experience, whether or not management recognizes this fact.

If a company can make a profit building a job twice the size of its previously largest job is, at best, unknown and, at worst, unlikely. Making a profit on a job three times greater than the largest previous project is actually close to impossible. How can an organization lacking experience with a project of such magnitude even determine what additional resources might be needed let alone what they will cost? Organizations that construct \$1 million or \$100 million projects, work in an altogether different environment than those that construct \$3 million or \$300 million projects.

Consider this example: A firm has two or three major jobs at a time, each around \$600,000 or \$800,000, while also working on a number of smaller jobs. The largest project ever was \$1 million and annual sales have been \$4 million with a comfortable profit margin. Then available work decreases, and the company's backlog drops off considerably. In desperation, the company goes after it a \$3 million project and captures it. The company thinks its problems are over for a while. In reality, their problems are just beginning.

No one in the company has considered the impact that a much larger project might have on the organization. Previous projects took a year or less to complete, and the firm's large projects typically started about the time another large project was about to finish, and there was usually another good-size project about half completed. Cash flow had not previously been a problem because when a large project neared completion, a large project in its middle stage was generating substantial monthly payments. At the same time, another large project was starting up and about to produce significant cash flow through front loading. This pattern is common in the construction industry. It develops because time and resources are required to capture large projects. Therefore, most organizations are limited to pursuing their larger projects one at a time which causes them to spread projects out comfortably.

Contrast this firm's normal "balanced" cash flow with what happens when they take on the $3 million job. At first, the front load is terrific, but the retainage adds up quickly and within months their accounts receivable will be higher than ever before. At the midpoint and toward the end of the $3 million project, accounts receivable will be staggering, potentially strangling the business. Finalizing a project of this size is a new experience, as is waiting for final payment. First-time larger projects are full of surprises such as the level of inspection and supervision by the designer's representative(s) because larger projects are usually managed by more-senior inspectors than smaller projects. Municipal, state, and lender inspections on large projects often create more recordkeeping and reporting than anticipated by firms with no experience with that size project. Occupational Safety and Health Administration (OSHA) and union work rules are usually more strictly monitored on larger jobs and security and safety requirements, and owner expectations expand with project size.

Size of project is a parameter of a construction organization's experience and forms their expectations particularly when the project is similar to the firm's experience in all other respects. The organization may be able to get the job done, but making a profit at it is another story and far from automatic. The situation is like a company that paves driveways and parking lots trying to pave an interstate highway project—similar but certainly not the same.

Changes in Types of Construction

For a variety of reasons, construction organizations expand the types of work they do, going beyond their proven track record. A few examples: Trying to add another type of work to their existing specialty, a highway firm took on a sewage treatment plant project and an earth-moving contractor bid on and won tunnel work. A building firm constructing one and two-story projects contracted for a high-rise project, and a firm that constructed office buildings took on a hospital project. These are all construction projects, but each requires different skill sets, expertise, and resources.

Contractors usually plan before taking on a new type of construction work, but too often drastically underestimate the cost of the learning curve. Some call this an entrance cost. They don't seem to anticipate the existence, let alone the cost of the inevitable learning interval. It is difficult to build work profitably with little or no prior experience and simply hiring someone who knows the new type of work is never enough. What inevitably happens is that an organization will struggle through one or more jobs before becoming able to execute the new type of construction profitably. The first-time costs can be extreme and the distraction from other projects almost caused by paying so much attention to the large job always adds to the deficit. Unfortunately, some companies do not survive the attempt.

Most contractors are much more specialized than they realize. Some bid on several types of projects but capture mostly one kind. They may call it coincidence, but it's more likely that they are better at estimating and constructing the type of projects they capture. Continually successful organizations usually start and remain with the types of construction they have experience in, and their growth and achievement are based on the continued refinement of that expertise. They perfect their estimating and execution and become more competitive, increasing their profitability.

Some readers will be asking where does the initial experience come from? The answer is that the experience comes from *very costly* trial and error. The failure rate of startup construction enterprises is extreme with most failing in the first five to seven years and success confirmed only after ten years in business. Ironically, most contractors that succeed past the ten-year mark are known for performing a certain type of work but seldom identify themselves as outstanding at that type of work. They, in effect, do not recognize their core competence and neglect to identify their proven specialty. Most "feel" that they can build anything and the longer they are successful at their unrecognized specialization, the more they believe they can build anything.

It is possible for a company to begin performing a new type of work without even realizing that they are making such a change. For instance, a more subtle change in the type of work involves changing from the public sector to the private sector or from the private sector to the public sector. This type of change has cost numerous firms a great deal of money because there are considerable differences between public and private work such as:

- Prebid qualification requirements
- The criteria used to select a contractor
- The amount of collaboration between the contractor, owner, and others
- The quality of work expected and delivered
- The number and cost of changes assumed to be allowed for in the bid or to be contracted for during construction (change orders)

A lack of knowledge about how the public and private sectors differ can result in constant and avoidable disputes resulting from differences in expectations. There can be significant differences in the construction process a contractor is familiar with despite the fact that the work itself is essentially the same.

Changes in Geographic Area

If a construction organization increases the size of their projects and expands into new types of projects, they will eventually outgrow their normal work area and will need to expand geographically. Depending on the size of the firm, their normal area of influence may be a county, a state, or several states. A construction company's normal area is the area in which the organization has a proven track record of profitability. Research confirms that almost any change in organizational operations will include risks. These risks must be recognized and planned for. Again, the question is not so much can the company build a similar project in a different location, but whether the company can build it for a profit? An organization becomes comfortable in their normal area; they may assume that they can perform that type of work in the exact same manner in a new area. Contractors may not realize that different areas they are unfamiliar with often have different customs, procedures, regulations, labor conditions, and more. The reality is work is done differently in different areas. In some cases, it is done just a little differently. The problem is profit can also be described as "just a little" so there is literally no room to maneuver when work is a little different than estimated or expected.

Profitable Project Selection

Very specific guidelines for profitable project selection were discovered and developed through years of research. Growth limits were announced more than two decades ago, yet the construction business failure rate continues. Most construction professionals seem to dismiss the risks associated with size, type, and location of projects. Some reject that safe growth limits even exist. Years after discovering and publishing the common causes of construction business failures, which are the foundation for growth limits, I conducted inquiries into why many refused to accept the limits.

The non-acceptance or disbelief in the causes of failure and growth limits seems to arise because many contractors see them as anti-growth. Avoiding the causes of failure slows growth and the drive for growth apparently outweighs the risks. Because the evidence that supports the risks associated with changes in size, type, and location of projects is so fundamentally logical, I concluded that there must be a strong motivator against the logic.

I found that because a common element of construction business failure when undertaken did not cause failure 100% of the time, many construction professionals concluded that the causes were not actually that risky. The reality is that changes in size, type, or location of projects may not always cause failure or may only reduce profits, so many decided that they must not be that dangerous. Because the failure does not occur 100% of the time, does not mean the research is in error, but for those driven by growth, it does. The research is conclusive. In most cases involving losses and business failures, changes in size, type, or location is the reason. Because loss or failure does not always occur, it does not in any way imply the *risk* did not exist anymore than surviving one pull of the trigger in Russian roulette implies that a risk to one's life did not exist.

Measuring risk is not accomplished by asking the question *did the event (or penalty) occur?* Measuring risk requires knowing the magnitude of the risk and its likelihood of occurring. For example, if you bet a dollar on the flip of a coin, the odds or risk of winning or losing are 50/50. If you bet one dollar, then the *magnitude* of the risk of winning or losing is one dollar.

Figure 2.2 Engaging in a common cause of business failure is like playing Russian roulette with your business.

The measurement of this risk would be described by anyone who can afford to lose a dollar as a *low* or *no risk* event. If you leave the likelihood of the occurrence at 50/50, and change only the penalty of the occurrence, does it change the measurement of the risk? Of course it does. And that is because the measurement of any risk includes both the odds of occurrence happening and the magnitude of the occurrence. If the bet on the flip of a coin is changed to one million dollars. Do the 50/50 odds now seem to be low risk? The odds have not changed but the magnitude of the bet (the penalty of occurrence) has and that changes the measurement of the risk.

Let's consider odds of six to one. Would you take a bet if the odds of you winning were six to one in your favor. Some may answer *of course* to that question, but you can't evaluate any risk just knowing the odds—you also need to know the penalty. Consider taking a project twice the size of any your firm has ever completed, or of a different type of project that ever attempted, or in different location than ever before worked in. This risk can be compared to taking a six-shot revolver, putting one bullet in the cylinder, and pointing it at your business. Is there risk before pulling the trigger? If you pull the trigger and the gun does not fire, does that mean there is *no risk*? Of course not. The odds of six to one were pretty good; however, the penalty is quite severe. If this potential project threatens the survival of your business or bankruptcy, would any reasonable businessperson ever accept that risk? I hope we can all agree—of course not.

The reason the common elements of business failure need to be taken seriously is that the consequences are lethal to your business. The reality that they do not cause loss or failure 100% of the time should not lead to *why not try it?* It means *don't do it* unless you want to put your entire business at risk. Hopefully, the Russian roulette example of six to one odds has not caused any readers to equate that to the construction business failure rate. The odds of business failure are much-much worse (Figure 2.2).

Summary

As the title of this chapter indicates, the motivator addressed here is ego, which seems to be the only explanation for the industry's obsession with growth. Psychologists describe ego as a powerful motivator that is inherently neither good nor bad. However, ego often has a negative connotation. In a positive sense, ego is described as healthy self-worth that can be negative if overdone. Self-confidence is common, even necessary in entrepreneurs. However, too much of it can lead to over confidence and too much of anything is generally not advised. I genuinely hope that this chapter convinces many industry professionals of the serious risks associated with growth, particularly too fast or too much growth.

3 Accounting for Work in Progress Is Almost a Joke

Introduction

In the modern construction industry, putting construction in place for profit is difficult. Accounting for construction project profit is even more difficult. The processes and procedures are complex to the point that many construction professionals complain they are confusing at best. Historically, there have been errors and misjudgments in reporting costs incurred in the field that have caused some to question interim profit projections. Others have described profit projections of work in progress accounting as *a joke*. This of course is not funny, and I have yet to find anyone who has said these things to who will agree to be quoted. Having unraveled the financial statements of hundreds of failed construction enterprises I tend to agree with those anonymous claims. If it was not so tragic, it would be tempting to describe financial statements profit reporting, at least for failed construction companies as a joke.

Some readers may consider parts of this chapter (or the entire book) *belly aching* from a disgruntled contractor. I am far from disgruntled. Having spent a lifetime in construction as contractor and after retirement, as an educator and consultant, I could not be prouder of our industry or more grateful for the life and opportunities it has provided. My books were an effort to give something back and this fifth and final book is a determination to openly discuss the closely held *secrets* those of us in the business for a long time have encountered. Ambiguities about some of our industry's shortcomings, inconsistencies, and things those of us devoted to the industry choose not to talk about. We produce the entire built environment with integrity and quality, and on time, on budget. We also manage an industry with the second highest business failure rate in the nation, so some things need to change.

Hard and Soft Numbers

Accounting for work in progress in construction is complicated and difficult. Work in progress is important in that the amount of revenue and therefore profit or loss in any year-end financial statement usually equals or exceeds the profit amounts from completed projects. This is significant because accounting for the profit or loss on a completed project can be accurate because it is based on a final accounting of all costs and expenses, whereas the cost to complete work in progress is based on an *estimate* of the cost of the work remaining. For completed projects it is an accounting function but for work in progress the cost to complete is an *estimate* by a project manager or estimator. The estimate is not therefore a *guaranteed* amount and is what some refer to as a *soft* (not exact or rounded off) number.

For readers not familiar with current industry methods of accounting for work in progress, a review may be in order. The value or dollar amount of the work yet to be completed is calculated

DOI: 10.1201/9781003500315-4

Figure 3.1 Accounting for work in progress relies on both hard costs and soft estimates that can distort reported profit.

by adding (the costs incurred to date) plus (the estimated cost to complete the project) and subtracting it from (the amount of the contract to date including executed and signed change orders without overhead and profit). As these amounts are taken directly from the contractor's books and records, one might consider them *hard* or accurate numbers. However, if unsigned change orders are in progress and only signed change orders can be included in invoicing to the owner, the contract to date amount may be questionable. The contract to date without the unsigned but potential change orders, would be understated, which would make it a *soft* number. Even if this were a hard number, the cost-to-date number may still be soft because subcontractors and suppliers do not consistently invoice on time. Some would suggest that the recorded costs on a contractor's books are always a little late and therefore consistently understated. Even if we assume that the cost to date is a hard number, the *estimated* cost to complete is then used to calculate the profit to date and the total cost to complete the project (Figure 3.1). These numbers are then listed in the financial statements with two decimal places of precision. Listed with two decimal places and not identified as *estimated* suggests these as *hard* numbers. My problem is that any number resulting from a calculation that included an *estimated (soft)* number is an estimated *soft* number. It cannot be a *hard* number.

Retainage

After signing a contract and before a project begins, contractors are required to prepare and submit a *schedule of values* for approval. This is a breakdown of the estimated cost for each category of work, sometimes referred to as *line-item costs*. These are assumed to be taken from the contractor's estimate for each category of work. For a large project the schedule of values will be hundreds of line-item costs and with the inclusion of amounts for overhead and profit, will equal the total contract amount. As construction progresses the contractor invoices monthly for work categories, partially or totally completed in amounts corresponding to the schedule of values. If a line item of work is fully completed, then 100% of the amount listed on the schedule of values can be invoiced. Similarly, if an item of work is 40% completed then 40% can be invoiced. The amounts actually paid will have had any retainage provided for in the contract deducted from them.

Construction contracts normally provide for the designer of the project to approve (or disapprove) the contractor's invoices, after which the owner pays the approved amount to the contractor, less retainage. Retaining 10% of the amount approved was historically common and still used on some projects. Retainage amounts have been coming down over the last decade with some projects reduced to 5%, some 10% only to 50% complete, and various other configurations. There does not seem to be any data available on the percentage of projects that remain at 10%. Because all public projects and most large private projects require 100% payment and performance bonds, which financially protect the project owners, many consider any retainage withheld to be excessive. Some claim that retainage is unwarranted and is actually *double protection* (retainage and bonds) and is punitive to contractors, referring to the practice as *belt and suspenders*.

Front Loading

It is not uncommon for contractors to list early work items on the schedule of values with amounts larger than originally estimated. For example, on a building project early items such as excavation, footings, etc. may be listed as higher than estimated amounts. To compensate for these higher amounts the contractor would lower line item amounts for work that will take place toward the end of the project such as painting, landscaping, etc. The practice is referred to as *front loading*. The purpose is to increase the cash flow to compensate for delays in the processing of pay requests and retainage amounts that are not received until the entire project is completed and accepted by the designer and project owner. Front loading is defended as an attempt to avoid having to pay interest on the retained amounts not received because the *costs to date* are real costs that are difficult to pay with retainage withheld.

There is an industry preference to maximize billing for work put in place. The practice of *front-loading* monthly project invoices has the effect of drawing money out of a project faster in the first half of a project than in the second half. Because there are generally projects in various stages of completion, industry wide front loading seems to keep contractors slightly ahead of billed amounts. On a cash basis the opposite is true with the construction industry unwillingly, and for the most part unknowingly, financing the owner's construction put in place. Some studies show owner's payments as received on average 70 plus days after invoicing which amounts to interest-free financing. Some readers may wonder how such a one-sided arrangement exists, and the answer is that it has been going on for so long that new entrants to the industry seem to accept as the norm.

On contractors' financial statements, front-loading has a huge impact, in that billing is immediately converted to an asset, accounts receivable. The net result is that most construction firms have invoiced more than their actual cost to date so that their financial statements indicate greater earnings than reality. While calculations of over and underbilling within the financial statements are supposed to compensate for this, it is fair to say that receivable are generally accurate or overstated. Collective accounts receivable industry wide are certainly not understated.

Weaknesses in Costs to Complete

This is not a complaint about the state of accounting for work in process and I do not have a better accounting process to suggest. What I am trying to explain is that we, construction professionals, need to understand the strengths and weaknesses in our own processes and understand that our accounting for the costs to complete our work in progress are far more impactful than most realize. Accurate costs to complete are absolutely critical to our success and survival. We have to upgrade some of what we are doing and maximize the effort, energy, and expense we put into managing the processes that produce the estimated costs to complete both large and small projects.

Almost every construction professional in the hundreds of construction operations that I have been involved in has told me that their people do an estimate of the costs to complete each of their projects every month. And most of the accounting staff in these firms also believe this. I genuinely hope that readers know better. My experience is that the column labeled *estimate cost to complete* on the work in progress report of the majority of industry financial statements is not an actual estimate of any kind. It is a calculation. More specifically, a subtraction of the amount in the column cost to date from the amount in the column contract amount to date. This is bad enough, but it will be casting doubt about the accuracy of both cost-to-date and contract value-to-date on many financial reports. When I speak with people responsible for the calculation of the *cost to complete* amount, most tell me something like: They are not estimators; Why would anyone do another estimate than the one that won the project; Everyone is too busy to do an estimate; and so on. One person suggested that to fix the problem the word *estimated* should be removed from the column.

A few claimed to do an estimate once or twice in the life of a project but when pursued further, it turned out they just reviewed cost reports or discussed the numbers with the project manager or superintendent. Not one person was able to produce an *actual estimate* or notes about their discussion. Some claimed they did not save it, and others admitted it was just a quick and dirty review of major items as a type of cross check. (I can't help but think about what to call that activity? Perhaps *a cross check of a meaningless review*.)

No Joke

When I say our accounting for work in progress is almost a joke, it sounds like I am making light of it. I might accept that if it was not such a serious and critical point. The reality is that the cost to complete the work in progress directly impacts the profit or loss reported on the firm's audited annual financial statements. And of course, everyone trusts those statements and relies on them! If the statements indicate profit, we continue business as usual. Even if the statements are wrong and the profit is less or indicates a loss, we continue business as usual. This is exactly what happened in all of the thousands of failed construction companies that I handled for bonding companies over the decade that I was in that business.

Lossing money is not the worst thing
that can happen to a contractor.

Lossing money and not knowing it
is the worst thing that can happen to a contractor.

Other Soft Numbers

While we are on the topic, we need to discuss some of the other *soft* numbers that find their way into our schedule of values and directly into our financial statements. These are numbers from construction business investigations that were discovered to be inaccurate (soft numbers). These numbers were not altered by anyone (at least not often). The inputs and processes produced inaccurate results that seriously impacted profit, loss, and value of the company.

Accounts Payable Timing

Accounts payable, the amounts owed to others such as subcontractors, suppliers, insurance premiums, etc. are in almost all construction companies understated for a number of different reasons. Let's start with the simplest reason, the accuracy of which any construction professional can check with their own accounting department to verify. Every accounting department, for very practical reason, needs to select a day/date of the month on which they will close the company's accounting books to produce the accounting reports and statements for the prior month. In theory they could collect data continuously, but you have to cease collecting data, referred to as closing the books, before producing the monthly reports and statements necessary to evaluate performance and run the business.

Your accounting department will verify that occasionally after the books are closed, an invoice for costs incurred in the prior month will arrive, typically in the mail. The accounting department cannot post that cost into the prior month's ledger because the books have been closed, so it must be posted in the current month. In doing so it will appear as a cost in the current month, not as a cost in the prior month when it was actually incurred. This has the effect of reducing the prior month's costs, thereby increasing the prior month's reported profits. The books and reported profit are in error, but the accounting department has no choice because this is how in how the late invoices are supposed to be managed under Generally Accepted Accounting Principles.

If invoices posted in the wrong month are a minor amount, the books and the reported profit are a little wrong. If the invoices are a significant amount, the books and reported profit are a lot wrong. I am not suggesting there is anything we can do about the incorrect profit being reported, but I am saying that the operations side of the business, project managers, etc., must be aware that this happens so they can understand how to interpret the financial reports they receive them.

There's more. For a verity of reasons some subcontractors and vendors do not invoice monthly or do not invoice at month's end, so the accounting department has no record that a corporate liability (a cost) has been incurred in that month. That cost will be missing on the cost reports that the field receives. This sometimes happens when a subcontractor, that needs only four or six weeks to complete their work from start to finish, elects to invoice only once when the work is completed. Therefore, if they perform the work three weeks in one month and three weeks in the next before sending an invoice there will be no record of the cost in the first month. There will also be no record of the costs in the second month if the invoice for the work arrives after the books are closed.

In this case profits are incorrectly overstated in both of the months that the liability/costs were incurred. The total costs for all of the work would then be recorded in a month when no work was completed, understating profits for that month. I am not suggesting the reports be disregarded altogether. But I am stating that cost reports should always be considered the least costs that were incurred. Some try to say that there was a plus and minus when sometimes the cost are more and sometimes less so it averages out. That is meaningless. If profits are down, you can take action. If they all of sudden go up, it is unlikely accurate. The field needs to know monthly where they stand and if there is any error, it always needs to be assumed to reduce profits, so action is warranted. If the information is better than reported, no harm. If you assume profits are fine and you take no action if the assumption is wrong, then it's too late. If the reporting is questionable, the only response is to discount the cost number and take action.

Unrecorded Accounts Payable

There is another reason some subcontractors and suppliers delay invoicing. The following is a process used mostly by smaller subs and suppliers. In an effort to defer income tax payment on earnings, some subcontractors and vendors do not invoice for their work or products in the last month or two of their fiscal year, particularly if they are experiencing a particularly high profit year. They then invoice in the first month of the next fiscal year, and when payment is eventually received, they record it as earned in that year, not when the work was actually performed or the product was delivered. This delays the payment of taxes on the income associated with the work in question for a year. Having done that, these subcontractors and vendors would have particularly high income and taxes in that year, so the process is usually continued in subsequent years. The net effect on the companies not invoiced in those months is to reduce their costs recorded and exaggerate their profit. As mentioned, this is common with mostly smaller subcontractors and suppliers so it likely to go unnoticed by the firm not invoiced and, if it is noticed, it possibly does not raise much concern because it improves their cash flow.

Some have said as long as the costs are eventually recorded it works out because in the months when the costs in question should have been recorded there were probably some other costs recorded incorrectly from the previous month. That statement makes me wonder how everyone connected with our industry's accounting process is constantly saying that industry accounting is 100% reliable to two decimal places. I am not saying the mathematics are in error. I am saying that sometimes the results are not describing the reality.

There were some confusing internal accounting processes in place in hundreds of construction companies that my firm investigated for sureties for more than a decade. The processes generated unreliable information and sometimes delayed information beyond the month in which it occurred and should have been recorded in. Most of the processes presented here were initiated by firms that were experiencing what they thought were minor cash flow problems or small losses. The processes seem to have been put in place to deal with what the firm's believed to be short-term problems. It seems as if these processes were not originally initiated to deceive anyone, but as the problems persisted the processes continued to mask cash flow difficulties and/or losses.

Checks Deliberately Delayed

Some contractors, in an effort to pay subcontractors and suppliers as soon as practical would have the checks made out to pay invoices absent the funds in the bank to cover them. The intention was originally to expedite payments so as soon as a project owner's monthly payment was received, the checks which were already addressed and put in envelopes with postage would be mailed. Unfortunately, processing a check in any bookkeeping systems automatically expunges the associated account payable. So as far as the books are concerned, once the check is made out, the account payable is paid and the amount of that liability no longer exists, which of course is not the case because there are not enough funds in the bank to cover the check. The expunged payable automatically but inaccurately increases profit amounts until the check is covered.

The problem gets worse if and when the anticipated payment that was going to cover the outgoing checks does not arrive or arrives at a reduced amount, which is not uncommon. If a designer decides that the contractor's requisition for payment is too high for any reason, a lower amount is approved for payment. As the unmailed checks are usually for multiple projects it becomes arbitrary who is to get paid, so some checks remain unmailed. If the problem persists for any length of time, the understated accounts payable amount grows, and all profit or loss reports continue to be inaccurate.

To avoid any accidental mailing of the unfunded checks, they are usually kept in a locked desk or filing cabinet. As desks and filling cabinets used to be made of wood, this process of holding check was referred to being *cast in wood*. It may seem hard to believe but this process can go on for a long time in one very large (huge) construction my firm was investigating the accounting department had developed a software program to manage the literally thousands of large and small checks that were made out whenever they were due. These checks would be recorded in an *off-the-books* program, stored alphabetically in a locked filing cabinet.

The checks would be retrieved and mailed by one of the few senior accountants that had knowledge of this practice when the funds became available to cover the checks. Available funds from all projects and sources were commingled and each day the software program would print out the longest overdue payments to assist in the decision making of which to mail. This information was combined with recipients who had complained the loudest or threatened legal action. When this construction company became insolvent there were well over a million dollars in checks locked in the filing cabinets (*cast in wood*).

Delayed Recording of Accounts Payables

Another practice that reduces accounts payable is recording invoices only after the amount of the invoice is approved by the project manager, field superintendent, or someone who has firsthand knowledge that the work was accomplished and acceptable or that the product was received as ordered. Many have tried to convince me that this is an appropriate process. However, this process involves multiple steps, and can take some time before a yet to be recorded invoice gets recorded because the person that approves or disapproves the invoice often works in the field or needs to check with someone in the field. The reality is that an invoice is a *liability* and moment an invoice is received by a firm and good accounting practice dictates that liability needs to be recorded.

An invoice is not a *request* for payment; it is a notice of claim in the amount of the invoice. A senior office of the corporation (not a clerk) may deny responsibility or refuse payment, but they cannot deny having received the invoice. Accounting practice and common sense acknowledge that it is a *claim* against the corporation (including lien rights) and refusing to record it does not change that. Some argue that the amount of the invoice may be incorrect and claiming too much. And that is a possibility. It may also be correct. That is not the accounting department's decision to make. They may think that sending it to others within the firm is not denying the claim, but it is refusing to acknowledge the claim.

Doing the above interferes with evaluation and management of construction operations. It renders the company books and records unable to the answer the critical question of, *are we currently producing our current work at a profit or loss*? Any delay in the recording of costs incurred weakens the management of operations. A delay in recording costs for the convenience or even accuracy of the accounting that negatively impacts operations can never be justified. The length of an avoidable delay, short or long, is irrelevant. It violates the primary purpose of all for-profit businesses: to maximize return on investment.

Case Study

Let's look at the exact process one rather large company put in place in order to avoid ever recording an invoice that was not 100% acceptable and approved by field personnel.

- Mail opened by a clerk, invoices placed in an outbox marked accounting
- An accounting clerk sorts invoices by project placing them in PM's inboxes

- PMs pick up inbox when in the often, in some cases once a week
- PM takes invoices to field, leaving them for superintendent's approval
- Superintendent returns approved or disapproved invoices at next PM visit
- PM negotiates disapproved invoices with vendor or subcontractor
- PM places approved invoices in accounts payable inbox when next at office
- Accounts payables clerk records invoices as accounts payable twice weekly
- CFO reviews payables weekly determining which to pay
- Accounts payable clerk processes invoice payment once a week

I am not going to suggest that readers try to determine what the fastest invoice turnaround is before it is recorded. The result here was that many invoices in this firm, to the tune of hundreds of thousands of dollars, were not recorded in the month the work was accomplished, or the month the products were delivered. The monthly profit and loss statements and the annual audited financial statements that this construction organization relied on were *almost a joke*. The monthly costs were continuously understated resulting in job profit being continuously overstated. All job costs were finally recorded but long after the project was completed. This company unfortunately failed, and their work was completed by their surety.

Numerous contractors and accountants defend this invoice approval practice with the argument that some subcontractors and suppliers overbill and we (the contractor or accountant) don't want our books cluttered with improper invoices. Therefore, we only record invoices that are verified and approved by our people. Some added, *you would be surprised how many invoices we get back from the field with reduced amounts*. The problem with this argument is that for a construction organization one of the primary uses of the accounts payable numbers are to measure performance. They are critical to measuring project by project profit or lose monthly.

Some readers may be thinking that this delay in approval process saves the company from paying inflated invoices. This is almost comical because recording of invoices has nothing to do with paying an invoice. All invoices are paid long after first recorded and never before being approved by the people approving the above invoices discussed. This discussion is about timing, not accuracy of payment.

All the process above accomplishes is that it delays the recording of invoices so the accounting department does not need to record any invoices that they may later find need to be changed. Accounting process never allows for a recorded number to be altered. If an invoice amount is recorded and later an approval process reduces the amount, the accounting process requires a new negative amount be added the book at a later date, in effect adjusting the amount. In this case the job costs were delayed and therefore incorrectly understated continuously to save the accounting department from any future adjustments. Once again this would be comical if it did not happen so often with disastrous results.

This bears repeating because it is happening out there. When I demonstrate within an organization where the principals had already told me it could not possibly be happening in their company there was a lot of shock. An invoice received is a claim against the firm and a potential lean right against a project. Sending invoices received around the company for any reason without first recording the invoice is in effect a *refusal* to recognize the claim. Not recording the invoice immediately upon receipt makes no business sense. The fact that a $10,000 invoice comes back from the field marked as $9,500 and is then recorded does not change the fact that someone in the company had knowledge of a $10,000 claim against the company and in effect ignored the claim awaiting verification. As an aside: No one in the firm has the legal right to arbitrarily deny any claim by refusing to record it. As a matter of fact, when it is finally recorded as $9,500 the $10,000 *claim* still exists.

Construction companies live and die on the timely and accurate assessment of the profitability of work in progress. If the records indicate profit, then field management has no reason to change anything because they are doing just fine. If the records indicate profit, but there is actually loss, then field management has no reason to change anything because they are doing just fine. In spite of this, some will still argue that recording the disputed amount would have been an error because the recorded amount would be inaccurate. This argument ignores the reality that the majority of invoices come back approved and unchanged. All the approval process accomplishes is to delay recording costs and to permanently exaggerate profit. This explains why so many project losses are not discovered until the project is completed. By then everyone that could have done something about turning the loss around is on other projects; hopefully not receiving positive, but fictitious profit reports.

Rare but It Happens

The last example of accounts payable issues is admittedly rare but encountered enough to warrant mentioning. There are some construction enterprises, particularly in smaller towns, that are among the larger customers of the bank they deal with. As such they are well treated.

Some contractors, promised payment by a client, will on occasion pay a subcontractor or supplier relying on the client's payment to cover the payment. If the client does not pay as anticipated, the contractor's account becomes overdrawn at the bank. For most people the check would not be covered (paid) by the bank and it would *bounce,* as this is commonly referred to. For favored or trusted bank customers, overdraws are sometimes covered usually for a small *overdraft* charge.

As mentioned earlier, the account payable was expunged when the check was written. In this case the check clears the bank without there being any reduction in cash (asset) in the contractor's books and records. There is also no reduction in accounts receivable because the payment from the owner that was going to cover the check hasn't arrived and therefore remains a receivable. The net result in the contractor's books and records is an increase in profit that does not actually exist. Some will argue that auditors would easily discover this; however, if the overdraft is covered by a deposit before the end of the month, the overdraft may not be obvious without a great deal of further investigation. In this case the bank never formally listed the overdrafts. (This is weak banking practice at best and the reasons why this and similar events are not discovered is not part of this text.) Suffice to say, this occasionally occurs and when it does it renders financial reports inaccurate.

Overstated Accounts Receivable

It may shock some readers to learn that accounts receivables were found to be overstated in both underperforming and successful construction companies. In successful firms, overstating of receivables was usually not by a large amount, but such overstatement is, nonetheless, an overstatement. Similar to our discussion about accounts payable, if some record keeping phenomenon overstated receivables and some understated them it would be less of an issue, but understatements of receivables were never discovered.

There is one exception. Some smaller companies deliberately understated receivables toward the end of their fiscal year to postpone income taxes. This pushes both the income and profit from the delayed recording of the payment into the following fiscal year, postponing the resulting taxes for a year. This distorts the records for both of the months where income is pushed forward and for the months in which that income is erroneously reported. As mentioned, this understatement of receivables is observed almost exclusively in smaller firms so industry wide it has minimal impact.

The most common overstatement of receivables is neglecting to write-off aged receivables that may be uncollectable. The fact that this happens is understandable in that the money has been earned, recorded, and in many cases, taxes paid on the recorded earnings. The motivation to collect funds owed and a contractor's natural entitlement to the money earned can result in false confidence that the payment will eventually be made, even if it requires legal action. Unfortunately, the age of receivables still on the books is often measured in years not days and experience confirm that the older the receivable, the less likely it will be collected.

Pruning Aged Receivables

Construction industry payment disputes do not always result in positive outcomes. The amount eventually received for disputed receivables is generally compromised (reduced) over time and the older the claim the greater the compromise until it approaches zero. In my experience, the value of most receivables older than a year approaches zero, if it's not already there. Those in legal action have only a limited chance of collecting enough to cover the legal expenses.

A closely guarded industry secret is that credit grantors discount aged receivables; with most over 120 days valued at zero. The reason for this is that credit grantors' experience is that most of these are eventually written off. The number of aged receivables on the books of successful contractors, while not usually representing a substantial amount, should be more carefully managed and written off if there is a possibility that they may not be paid. That does not mean all efforts should dropped to collect them. It means that overstating overall profitability is dangerous. Reporting even slightly inflated profit tends to inflate optimism about performance. It can also affect performance bonuses paid out to employees.

It should come as no surprise that aged receivables within underperforming or failing construction enterprises are very often a substantial amount. Obviously if profits are suffering, the pressure to leave potentially uncollectible receivables on the books is huge. Experience confirms that the actual age of receivables is directly proportional to the likelihood that they will be collected in full or at all. It is critically important to have a policy in place that scrutinizes the collectability of aging receivables and that encourages writing off questionable receivables in a timely manner. The rule of thumb is not: *We once collected a receivable that was a year old.* It is more like when a receivable is six months old we write it off. If any one of those is eventually collected it is recorded as *luck*. Exaggerated profits detract from performance. Understated profits do not. The choice is simple.

Real Risk

Financial record keeping is critical because it can cover up as well as expose real risks. Managing payables and receivables should concern all construction professionals because they entail huge risk. We need to begin to cause change. My greatest fear is that readers will be thinking the accounting and record keeping shortcomings addressed in this chapter are rare or are exceptions. They are not. While some of the evidence discussed in this chapter was found in underperforming companies, it was found in performing companies as well. Remember that only companies 10 years in business or longer are in this database, which means that every one of the failed companies discussed were successful before they became unsuccessful. This database includes second, third, and fourth generation firms.

Reading Financial Reports (Accurately)

What the industry needs to do is to take steps to accelerate the internal cost reporting process and demand accurate and timely reporting. At the same time, we need to modify how we read financial reports and statements. We need to amend our thinking about line-item entries because we (readers) now know that:

Accounts payable in construction financial reporting are accurate or understated. Never overstated.

Therefore, profits are accurate or overstated. Readers also now know that:

Accounts receivable in construction financial reporting are accurate or overstated. Never understated.

Therefore, profits are accurate or overstated. Note that all of the record keeping anomalies presented here outstate profits. During the analysis of the literally tens of thousands of internal reporting systems and thousands of audited and unaudited financial statements, not a single anomaly was found that offset the overstated profits. We should take a breath to absorb what this really means. All the collections of data and annual reports on the performance (profits) of all of the different segments of the construction industry have been overstated. Some will say *it is just a little overstated.*

For the collective audience that receives this information it may appear harmless. However, accurate profit reporting for each individual business is critical. You can't manage a construction business if you can't definitively measure profit by activity by week (perhaps by day) and at the very least by month. If profits are overstated just a little, then perhaps just a little adjustment to operations is warranted. If profits are overstated a lot, then a lot of adjustment to operations is warranted (if it is not too late). If construction professionals refuse to understand and accept that if undisciplined industry processes continue to overstate profit, then nothing will change, and the obnoxious industry failure rate will continue.

Diminished Final Payments

The industry is plagued with a history of delayed or disputed pay requests pending designer approval and/or slow payments from owners to contractors once payment requisitions have been approved. Entire pay requests are sometimes withheld over a disagreement over the value of one small item or the acceptability of some piece of work. These often result in contractors agreeing to credit some amount to settle the disagreement artificially reducing a receivable. The more common difficulty is final payment, which in some cases does include the last two or three months of withheld payments. It is not uncommon to perform unreasonable or inappropriate punch list work in order to collect final payment, which almost always includes retainage for the entire project—a staggering amount.

Final payment seems to be the bane of all contractors' existence along with never-ending punch lists. Final inspections of projects never seem to be final. Some designers or owner's representatives seem to be impossible to satisfy with the third or fifth punch list having new items on them. I personally experienced a broken window in a school on the fifth punch issued six months after occupancy, which did not appear on punch lists 1 to 4. When that was brought up to the designer's inspector his reply was, *I don't care when it was broken. I am not accepting this building with a broken window in it.* The window was fixed at no cost to the owner. Why? The contractor was desperate for final payment including his retainage.

The economic pressure on contractors many months after a project is in use or occupied by the owner is enormous. Being right becomes less important than buying one's way out of reasonable demands over unreasonable periods of time or unreasonable demands over unreasonable periods. This type of thing happens too often and the financial impact on the contractor can be severe, including serious amounts of financial compromise. The net impact is an instantaneous reduction in accounts receivable on a contractor's financial statement that credit granters were relying on in their credit evaluations. Any erosion of accounts receivable tends to be suspected, and contractors' trustworthiness can be affected.

Construction Claims

The production of construction projects seems to generate more than its share of disputes, particularly on low bid lump sum contracts. Projects are designed by a third party who have financial motivations and constraints and owner/clients that often want one-of-a-kind designs that can be challenging. Owners and designers often want changes made during construction or have different expectations of what is actually designed or built. These things challenge relationships and often result in differences of opinions that can lead to disputes (Figure 3.2). Simple agreed upon changes to the project can result in disputes over the value or who is responsible for the cost. Changes are often completed before the cost details are finalized or change orders are issued, which eventually are reduced to claims.

Claims arise when contractors find that portions of a design are impossible or impractical to produce; that specified products are not available; and/or any number of misunderstandings or misinterpretations of the design occur. Claims are most common on lump sum projects where the

Figure 3.2 Disputes over final payments and claims often erode contractor profitability and strain financial stability.

contractor had to be the low bid to get the project and there was no *room* in their estimate to pay for any portions of the work not included in their original estimate.

Unrecovered costs claimed and the cost of pursuing claims are job costs that reduce profits or create loss. There is no specific category the costs fall into but for consistency, here they may be considered an unrecoverable account receivable. Most consider this a risk of doing business, but a solution is to not bid on projects where the designer is unknown or known for incomplete designs. The obvious response to this will be how can a contractor get experience with new designers when expanding? The answer is, very carefully and start with smaller projects. Gaining experience is a constraint to growth, which is why rapid growth is never advised.

Summary

Construction is obviously a high-risk business on the production side of the business, but this chapter and this book is restricted to risks on the business side of the business. These business risks are real and particularly insidious in the construction industry because most successful construction professionals have little or no formal business education. I often hear that *if we make money on the production side of the business everything else will take care of itself.* This book is proof positive that such statements are not true. It is the business side of the construction business that causes failures in an industry that ranks as the second most risky business in the country.

4 The Industry Has Abandoned Cost Control

The History of Cost Control

I started in the industry in 1960 as a carpenter apprentice. The term in those days was *apprentice boy*. Back then construction crews had the estimated unit costs in hand when they started a work activity. Knowing how we were performing was a target. The office already knew what the material and equipment costs were, so the only thing the field was interested in was the estimated labor costs. As soon as the work started the field knew how their production measured up. Certainly, before the office knew. The process was a natural motivator. The crews wanted to beat the estimated costs and over time they wanted to beat their own costs. Cost control systems were managed by the production/operations side of the business in cooperation with the estimators who provided the unit price information. The cost control process was not very *formal*, but it produced excellent results. The field knew weekly, and often even daily, if they were achieving or beating estimated costs. Field calculation might have been done on the back of an envelope and numbers were rounded off, but everyone knew how they were doing every step of the way.

Over time cost control procedures evolved. Stimulated by the advent of computers to assist with the calculations combined with a desire to formalize the cost control process, a variety of different methods and processes were tried. Eventually cost control became a very sophisticated process of data collection and the presentation of data. The process was moved into accounting departments and accuracy became paramount. The cost control processes were introduced into each construction company's accounting general ledger where it was required to be accurate to two decimal places. With the advancements in data processing capacity, new cost control software programs were being developed coast to coast. Job cost capabilities were typically add-on modules to accounting solutions which were largely focused on financial reporting. Software firms competed for the business of contractors by offering value in the number of ways cost information could be collected and presented.

As accounting professionals trained in generally accepted accounting principles (GAAP) began providing cost information to construction teams who needed the information to manage their work in the field, accuracy became the driver. Two decimal places of precision became standard, and timing was a problem. For readers unfamiliar with accounting processes, financial data is carefully and accurately collected. However, to use the data for reporting purposes, a cutoff date needs to be agreed upon. Accounting data has been historically reported monthly. Each firm has an agreed upon fiscal (financial) year which defines the calendar year for reporting purposes, but it can be any 12-month period. To close a firm's books monthly, a date in the following month must be agreed upon. That date would optimally be after most, if not all, of the payables (invoices) for the prior month had been received. For a smaller company the cutoff date might be as early as the 15th day of the following month, but for larger firms it could be the 20th day, 25th day, or later.

DOI: 10.1201/9781003500315-5

Cost Control Today

I am guessing that readers are sensing a problem with the evolution of cost control systems. Cost control was placed in the accounting department and existed in the company's general ledger. Current cost reports, generated by a firm's accounting system, were issued monthly. The term "monthly" is a little misleading because the cost reports cannot be issued until the company's books are closed. If the books close on the 15th of the following month, the cost reports for the prior month are issued six weeks after the work being reported in the cost reports occurred. Therefore, costs reported for work activities performed early in the prior month are six weeks old, and the work activity may have been finished by then. The reporting delay is even longer for companies that close their books on the 20th day, 25th day, or later in the month. The reality is that field management needs this *accurate* data sooner. If the report is six weeks old or more, it is reporting on work activities that are partially completed or finished already. This leaves the field-people no time to make corrections. The information is too late to be of any practical use in the field.

Reporting cost over runs is of no use to the production side of a business. It has value only if it is received in time to use it to take corrective actions. Cost information reported later than corrective action can effectively be taken would more correctly be defined as *cost measurement* instead of *cost control*. The late information is of no use to operations but may be of some use to the business side of the business as an accurate record of the loss incurred.

I am not complaining—I am reporting. The reason I began this chapter with a history lesson is an attempt to explain that new and innovative processes, developed for proper and logical reasons, may or may not work. Our industry sensed a need to improve existing cost control procedures to better manage the production of the work. In this case it did not work.

Historically the industry had cost control processes that were timely but not accurate to two decimal places.

What we replaced those cost control processes with provided more detailed data, that was believed to be more accurate, that could only be reported too late to be of any use, other than for record keeping.

The new processes were not developed by the *users of the processes* but by computer experts. The new processes evolved over two decades during which the critical need for timeliness was somehow misunderstood and the accuracy of the information became the primary driver. This is not uncommon in industrial development. The significance today for industry and construction professionals is our reaction to this condition. The current method of cost control with accounting solutions focused on financial reporting does not work. Maybe there is a method out there that can be both accurate and timely, but contractors cannot wait because the industry failure rate continues to rise at a staggering pace.

The ONLY prudent reaction is to return to the proven field-managed cost control methods that we know work.

Back to Basics

To begin with, *cost control* is a misleading term in that the process *controls* nothing. The process that we in the industry call cost control records what it costs us to produce a unit of work so that we can compare that to what we estimated that cost would be. If we do that accurately and at the very beginning of the work activity in question we can determine if we are producing the work for less or more than the estimated cost. If the actual cost is less, it pretty much universally causes us to do nothing. If the cost is considerably less, we might even celebrate, but that is rare because overestimating the cost of the work often results in pricing yourself too high to obtain the work. If the cost incurred matches the estimated cost, in practice, most do nothing. If the cost incurred is higher than estimated, in theory someone is supposed to take action. However, in current industry practice there are typically numerous reactions other than correcting the activity and fixing the problem.

The all-too-common reaction is for those managing the work to say that the activity in question has just started, and it always takes a little time to get organized. The pace will pick up in no time and we will be back on track. Another reaction is that *we don't have enough people on the crew, or enough crews, to be efficient*. As soon as we *gear up*, whatever that happens means to the person saying it, things will get better. The opposite reaction is also commonly stated which says we have too many people or crews, and as soon as we cut back things will be fine. Other reactions are: there are equipment, scaffolding, and rigging excuses; or they sent me the wrong size—too big, too small, etc. The weather is also a common excuse. Too hot, too cold, too wet, too dry. There are too many reactions (excuses) to list—lousy design, over inspection, poor skill level of the workers, union disputes, and so on. The worst excuse of all is, the estimate is wrong, and the work cannot be produced for the amount listed.

Blaming the Estimate

Construction companies estimate the work and sign a performance contract; and if they intend to stay in business for any length of time, they must complete the work on time and on budget. Continuously managing the cost of the work, including immediate detection of cost overruns, is fundamental to success. Current ineffective cost control processes, as described earlier, continue to support the industry's appalling failure rate. The delay in discovering and reporting cost overruns has somehow evolved into a reaction to investigate who is to blame for the overruns rather than taking immediate actions to correct the problem. This has the same effect as not reacting at all, which translates into continuing to lose money. A prudent construction professional takes immediate actions to fix the problem.

One firm used an old football line to describe their policy about what to do when production costs were higher than estimated. Their policy was to *stop, drop, and roll*. In football when the ball carrier is facing a defense blockade, rather than drive into it and risk losing the ball during the collision, the carrier is to stop, drop down and roll. Roll forward, clutching the ball securely to maintain possession of the ball. The primary objective changes from moving the ball forward to don't lose the ball. The rules change with little notice. The same thing happens in construction.

If any work activity, except a very short one, exceeds estimated costs by anything other than a tiny amount, the thing to do is stop the activity (stop, drop, and roll). You can't continue the losing activity because adjusting an ongoing activity is impractical, inefficient, and only supports continuing the activity for a loss. There are several arguments that support adjusting as you go, but experience confirms that continuing the losing activity does not work. Continuing forces adjusting the work process while it's in progress and precludes exploring major changes to the work process that may yield a better solution. Stopping the activity stops the loss and also provides time to evaluate the cause(s) and develop a fix. Or to determine if it is advisable to call upon the experience of other employees (or experts) rather than just those currently involved in the losing work, who may have a commitment to (or pride of ownership in) the current losing process. Let's look at a couple of examples to clarify this concept.

Intervention Processes

First Example: A self-performing general contractor in the west began the erection of the outside precast concrete skin of an eight-story building that included a huge crane and a crew of five skilled craftsmen. After several days of work, the cost of placing the precast panels was way above the estimated costs. The project manager, superintendent, and VP of construction were studying the problem, but were stumped. The work was stopped and an experienced constructor who was unfamiliar with the project was called upon. The constructor was unable to visit the site on short notice, but the current erection process was described to him in detail, and he was able to watch films of the actual erection that took place over the prior several days. He was able to study the work in progress films in slow motion as well as stop, rewind, and play back in slow motion.

The newcomer noted that the crane had to hold up each piece of precast until it was welded on both sides by one welder who then had to wait quite a while for the next panel to be hoisted into place. This delay was because the crane had to unhook from the precast piece that was welded into place, lower and hook to the next piece of precast on a truck, then hoist and be guided into position on the building to be welded. Also noted was the person hooking the lifting cables to the precast panels resting on a truck bed took a long time running from each side of the long pieces of precast to connect each end. That same person then had to get off the truck to handle the guide rope that kept the precast that was being lifted from swinging too much and assist in correctly aligning it as it was put into place on the building. It was noted that, of all the sequential steps that had to be taken, crane efficiency was the key. The cost per hour of the crane, fuel, and operator was the costliest element of the precast erection process.

The constructor made recommendations which were put into place the next day and the work was restarted. A second welder and welding machine were added in the building to expedite the release of the crane so it could go back to the truck for the next piece of precast more quickly. When they were using only one welder all six connecting points on one end of the large precast panel were welded before moving the welding machine to the other end of the precast panel and welding the opposite side. The second side of the precast was also completely welded in place before the crane would release the load. With the second welder, it was now possible to begin welding the connections on both sides simultaneously (Figure 4.1). The outsider also reminded the crew only one of the six weld plates on each side had to be welded before the crane could release the load.

Using two welders and two welding machines simultaneously, the connection was accomplished in minutes after the precast was hoisted into place. This compared to almost 40 minutes the prior method had taken. The other five weld plates per side were welded while the crane was picking up the next piece of precast, which allowed just enough time to move the two welders and machines to the next location. As per the recommendation, a second person was added at truck level so both sides of the precast panel could be hooked to the crane at the same time, cutting the crane time (delay) for that activity by more than half. A second guide rope and person were also added to eliminate the time to swing the unwieldy hanging precast into place on the building.

First Example Results: Prior to this intervention the crew had averaged setting four precast panels a day. After adding three people to the crew, a second welding machine, and another guide rope, the crew set seven precast panels on the first day, which happened to be the pre-bid estimated productivity. After a week they were averaging nine panels a day. The precast operation took more than three months and was completed at almost 19% under the estimated cost.

Second Example: A foundation subcontractor in the northeast had a lump sum subcontract to excavate and construct a six-level underground parking garage in a mid-size city. The top of the garage would be the first floor of a multi-story apartment building. The project site was a corner

Figure 4.1 Building panels being hoisted into position.

lot with very limited room to work. There were existing buildings on two sides of the site and city sidewalks and streets on the other two sides. The city allowed for the sidewalks and one lane of traffic to be closed for the duration of the project along both intersecting streets. All excavated material would of course have to be trucked away from the site. The contractor who got the foundation job had a cost advantage at bid time because he had found a site several miles away that needed dirt. He had a tentative deal to sell the waste dirt from this project as fill dirt to the second site. This seemed like a perfect combination until the project began.

The foundation contractor had problems from day one. When the job was ready to start there was less staging room for the empty trucks than he anticipated. Half of the closed traffic lanes were filled with the general contractors and designer's office trailers, which would eventually be stacked two and three high with the other major subcontractor's trailers (without wheels). That left only enough room for four huge dump trucks to be waiting to be loaded with excavated material (dirt). As the excavator was only able to load each truck one at a time, numerous other trucks had to drive around waiting for their turn in line because there were no other places to stage them. They certainly could never find places to park.

The contractor had planned to use 20 trucks for the round trip back and forth to the dump site. But on the first day after 20 trucks were loaded and rolling an empty truck did not return for almost two hours. They were all stuck in traffic traveling both ways (Figure 4.2). The contractor was able to pull trucks off other projects and engage a good number of independent truckers he regularly used. That doubled the number of trucks by the third day. However, the first empty truck was still unable to return to the loading site any faster. Because the trucks accelerated from a stop slower than the automobile traffic, all the extra trucks did was create more traffic. By late mornings the

Figure 4.2 Dump truck hauling in congested traffic conditions.

timing would improve just a little, but by late morning lunchtime congestion built up so much it created congestion right through the evening rush hour.

The progress schedule was in shambles with the general contractor demanding more trucks. A few more trucks were added with little improvement. The general contractor demanded longer hours, which the foundation contractor said were out of the question because there were no overtime costs (time and a half in this area) in his estimate. It looked as if a legal battle was coming.

Second Example Solution: The foundation contractor turned to a consultant who spent time at the site, studied the conditions and by the end of the day recommended the solutions. The union in that district allowed for shift work with compensation at eight hours pay for seven hours work. Referring to traffic patterns for the city in question, the consultant recommended that the excavation be conducted exclusively at night. With a 6 PM to 1 AM shift and a midnight to 7 AM shift using 20 trucks as originally planned. He advised reducing the number of trucks if there were long waiting times at the loading process, and adding trucks so there would never be a time when trucks were not being loaded. The consultant reminded the client that in excavation and earth transporting operations, continuous loading was generally the most important aspect. Loading should never be delayed or stopped, if at all possible, which means it is imperative to always have trucks ready and waiting to be loaded.

To accomplish this all maintenance and fueling of the excavator and trucks should be conducted during non-loading hours, which in this case means daytime hours. If that would not possible, he recommended a second, perhaps smaller, excavator should be available to cover any potential excavator downtime. He called this process *supporting the weak link* theory. The contractor adopted all the recommendations and with earth moving offsite continuously during the two night shifts they caught up with the schedule in a little over a week. The entire excavation was completed 34 days ahead of schedule and profit from that portion of the work was almost 40% more than originally estimated. A potential financial disaster was prevented by rethinking the entire process. And a night shift was being considered for concrete placing portion of the foundation and structure of the underground parking garage segment of the contract.

Lessons Learned

One of lessons here is those closely associated with an underperforming work activity may not necessarily be the best choice to investigate the causes and develop solutions. Those who have selected or are working on the current process are at a disadvantage because they are too close to the problem. For most people, they develop a certain amount of *pride in ownership of the current process*, which can cause them to believe that the process just needs to be *adjusted.* All attempts to fix an underperforming process must include considering entirely new processes. This is not intended to suggest that new processes are the only cures. In some cases, adjustments to the existing process may work, and while some underperforming processes will have no improvement or just some modest improvement, some may have spectacular results.

The purpose of the examples is to encourage diligent intervention as soon as any work item is discovered to be underperforming when compared to its estimated costs. Questioning the estimate is NEVER a solution. Once work is under contract, the budget for the work is established and it is the operation's responsibility to produce that work at a profit no matter what the estimated costs were. Saying *the estimated costs are wrong or it's not my fault* is the same as saying *It's not my job.* If it is estimating's job to get the work and they got this job. Then it has to be operation's job to do the work. So do it. It is difficult to estimate the work and difficult to produce the work and both departments make mistakes.

The reason I am pushing this issue so hard is that too many of the failed projects I have been involved with on behalf of sureties were underperforming projects that could have been corrected, but were left to fail. By that I mean that the construction professionals that discovered the underperformance spent their time trying to determine who caused the problem and pointing blame rather than attempting to fix the problem. I am hoping that readers understand that we are constructors not complainers. Construction is a tough, risky business. There are risks in estimating, administrating, sales, accounting, and operations so it's not hard to place blame; however, determining who caused a problem does not correct the problem—it is meaningless. Only operations can fix an underperforming project therefore it is operation's direct responsibility. Placing blame or complaining is dereliction of responsibility because it distracts from fixing the problem. *End of story.*

We should not have to be even discussing this topic. However, having been in the failure-business has taught some odd lessons that should be obvious but don't seem so to everyone. For example: everyone in an organization is on the same side, is responsible for the generation of profit, is responsible for minimizing loss, and is responsible for fixing problems within their area of responsibility. The successful construction professional is a team member who works hard, fixes problems, and realizes that there is no profit and nothing to be gained by placing blame.

The point is that cost control processes as they exist today in most construction companies do not work well. Somewhere along the way we traded timeliness for accuracy and gave up our ability to take corrective action when it matters most—during the work. Field-managed cost control used to be immediate, intuitive, and effective. Now it's buried in overly complex cost reports and arrives weeks too late to react to. That is not cost control. It is cost reporting. The solution isn't complicated. We need to go back to basics and give the responsibility for cost control back to the operations team, where it belongs. Immediate awareness of cost overruns is critical to allow for rapid intervention. Our field and office teams need to be focused on solving problems, not placing blame. It was the process before cost control was moved into the accounting process. The prior process worked and it's time we start doing what works again.

5 Getting Paid for Your Work

In the world of construction, completing the work is only half the battle—getting paid for it is the other half. Far too often, contractors find themselves burdened by payment delays, reduced invoices, or ambiguous approval processes that shift power unfairly into the hands of designers and owners. While contract documents appear to promise timely compensation for completed work, the reality on many projects is that contractors are left chasing payments they've rightfully earned. This chapter explores how the traditional contracting process undermines payment accountability and offers practical strategies to help contractors assert their rights, protect their cash flow, and uphold the integrity of their business.

Construction Contract Fundamentals

In the construction industry the design is generally produced by a designer who has been commissioned by the owner of the road, bridge, or building to be constructed. While there are some design/build projects, they are an extremely small percentage of the industry and they almost always magnify the construction firm's risk by adding the design risk. The classic construction contractual arrangement is a contract between the owner of the project to be constructed and the designer of the project—an architectural or engineering firm commissioned to design the project. And after complete or partial design, depending on the type of delivery method chosen for the project, the owner of the potential project hires a contractor to construct it. The types of contractual arrangements between the owner and contractor are too numerous to delineate. There is no contract between the contractor and designer; however, designers generally have some degree of authority over the contractor without any responsibility for the work of the contractor.

In lump sum contracts, most common up until the last several years, a completed project design would be bid on by multiple contractors. Each bidding contractor is, in theory, qualified to construct the project. Lump sum contracts still exist but are currently being supplanted by numerous attempts to come up with a better contracting method. Few in the design and construction industries seem to want to address the motivations for the continuous experimentation in procurement methods. An objective observer might suggest that the motivation for experimenting with a huge number of contracting methods seems to be a desire on the part of owners, designers, and contractors to avoid any liability for the quality of the design or the construction of a project.

Complicating this desire to avoid liability is the fact that there are three participants in the construction process (owner, designer, and contractor), but only two contracts: a contract between the owner and designer, and a contract between the owner and the contractor, with no contract between the designer and the contractor. This in spite of the fact that the two contracts grant the designer significant authority over the contractor. This authority comes through the designer's inspection and approval authority of virtually every element of the contractor's work, and over the contractor's

DOI: 10.1201/9781003500315-6

payments. The two contracts, while granting this authority to the designer, also protect the designer from and responsibility for their decisions about and directions to the contractor. These are genuinely complicated issues subject to the interpretation of two separate contracts involving three participants—only two of which have contractual relationship with another.

Project owners regularly claim that they have no part in the design or the construction of the project so they should not have any responsibilities for either. Designers point to language in their contract with the owner and the owner's contract with the contractor that give them authority over the contractor in many areas of the work. And then designers point to language in both contracts that state that the designer has no responsibility for the results of their authority over the contractor. The contractors have lost this debate and are contractually saddled with responsibility for their own actions and numerous implied responsibilities for the actions of the owner and designer. In effect the contractor works for both the owner and the designer who both have the right to inspect and approve the contractor's work and, perhaps more importantly, the contractor's payments.

The two-contract arrangement, intentionally or unintentionally, provides designers with considerable authority over the contractors in spite of the fact that no contract exists between these two parties. This *authority*, real or not, combined with the power to reduce or withhold payment requisitions has metastasized over time into contractors voluntarily yielding a certain amount of control over themselves to the designers. The control is demonstrated in the approval process where contractors often have to *jump through hoops* to get shop drawings for products specifically specified approved. And at the same time suffering production delays waiting for designer field inspections and approvals. This creates an environment that grants the designer considerable control over the contractor and the contractor's progress. Because the contractor's progress and success are dependent on timely approvals, inspections, etc., the designer has the upper hand in this relationship.

A minor demonstration of *unbalanced power* is the payment process. A common practice that has developed over time is walking a project with the designer (at a time convenient to the designer) to agree upon what percentage of the work is complete and acceptable to be invoiced for. Contracts usually say something like *invoices for work completed by the last day of the month to be paid on a specific date the following month*. The contract also says the owner will pay only what the designer approves. Many designers suggest contractors walk the project with them to agree on what is complete and acceptable, claiming it will save time. This tends to shift power because the designer is the final word in payment approval process, period (Figure 5.1). Many contractors defend this by saying it is the only way to ensure timely payment. That statement in and of itself confirms my position on who has the power in the arrangement. My position is that a contractor knows how much work is completed and needs no advice on the quantity of the work. Contractors also need no advice on the correctness or quality of the work in place as they are more than capable of measuring that.

A walk through is simply an agreement to negotiate what work is worthy of being invoiced. It is an agreement in advance that the designer will have the last word. You might as well suggest that the designer walk through and let me know what they will allow you to invoice. Contractors tell me they don't want to have to type the payment line items twice after the designer red lines the submitted invoice and demand that a new invoice be produced. The designers usually say *you don't want the owner to see that you tried to overbill, do you?* My answer to this is that it is my invoice, and the red lines are your opinion. Submit it as-is and I will deal with the owner.

Contract Payment Provisions

According to most contracts, the payment requisition is to be filled out by the contractor and approved by the designer. A contractor doesn't need any help determining what work was performed during a month nor how much they should be paid for it. If the designer doesn't want to approve it,

Figure 5.1 Construction contracts often grant designers authority over payments, leaving contractors with little recourse.

that's their prerogative. They can red-pencil it and send it on to the owner along with the amounts they think should be paid. There's no need to retype it any more than a need to walk the site and bargain for the amounts. All that establishes is that the designer, and only the designer, decides what the contractor will be paid. Retyping means changing the request to say what the designer says it has to say.

What contractors accomplish by these practices is to place their monthly payment amount totally in the hands of the designer without even a right of appeal. They can argue and cajole, but once they have agreed that only a clean, typed requisition can go to the owner, they place 100% of the decision about how much will be paid in the hands of the designer. When this happens, the contractor has voluntarily agreed that the designer's amount is correct. This is ridiculous because it not only rewrites the payment provision of the contract but also abandons *entitlement* to payment leaving contractors in the position of negotiating for payment. The reality is the designers don't have to pay the subcontractors and suppliers and haven't already paid for labor. They aren't under any time constraints at all.

So why do so many get involved in these practices? Some say *it's easier than arguing over payment and arguing will only cause other problems with the designer*. Others have said, *I don't want the owner to see a red-penciled requisition, because they'll think we were asking for more than we should have*. Still others said, *the owner's not going to pay anything except what the designer approves anyway, so what's the difference*? The difference is that the practice gives the designers a huge amount of control over the contractor during the construction process. The reality is that as long as a contractor intends to live up to their end of the contract (including every item in the plans and specs as most designers insist on anyway) they have nothing to fear from the designer.

There is no need to shelter the owner from payment requests. Owners have an active role in the construction process and one of the few things the contract says owners must do is to pay for the work performed. There's no harm in clarifying that and involving owners in the payment process right up front. If this sounds like I'm not in favor of cooperation be advised that I strongly recommend cooperation. I'm just in favor of cooperation on sensible and business-like terms. I am not in favor of one-sided cooperation where the contractor must keep quiet, so the owner is always happy, or where the designer's boat is *never rocked.* This is one of the excuses used during the payment process when contractors are advised not to *rock the boat by* standing up for their rights, some of the very few rights granted in the contract which are actually favorable to them.

Contractors are directed in the contract to requisition for work performed, and the owner is directed to pay within a specified period of time. If the designer doesn't approve the full payment request, the owner only must pay the approved amount. The contractor by requisitioning is simply doing what the contract calls for, as is the owner in paying only the approved amount. The designer, on the other hand, has stuck their neck out because if there are any consequences to the designer's action of reducing the payment, the owner will end up paying for it. If owners don't know that, professional contractors make sure they learn it. Unfortunately, too many contractors seem to think they are powerless in the payment process.

Payment Case Study

When I was a contractor, I returned a red-penciled invoice to a designer which had been sent back to me for retyping. I had it hand delivered back to the designer with a note advising that the requisition be send on to the owner red-penciled as it was because I was not going to be changing it. I also copied the owner with my note and the red-penciled invoice. I got a call from the by then incredulous designer saying that they had never, and never would, sent an owner anything but a clean typed requisition. This was the first requisition on the project, and I knew he had not visited the project for several days after receiving the invoice, I assumed he had reduced most of the quantities out of habit or perhaps because he thought he should. I knew for certain that the amounts invoiced were correct and some, understated quite deliberately because it was the first invoice. There was one line item where anyone could see that some work was obviously completed and that no payment was requested.

The designer insisted he wouldn't send the red-penciled copy to the owner, and I would just have to change it, retype it, and sign a new one. I politely told the designer that I never retyped any submitted invoice and advised him that I had sent the owner a copy of his red-penciled invoice along with a copy of his (the designer's) letter telling me to change invoice. I added that I was waiting to see what the owner wanted to do. The designer proceeded to inform me that the contract stated that I could have *no contact with the owner, except through the architect*. I responded that he would have to explain that to the owner. He then said to me, *what if I advise the owner not to pay you anything*? I made no reply because I assumed that he knew that he would have a hard time explaining that to the owner. It also sounded too much like discipline for me to even answer.

The designer followed through and immediately advised the owner to ignore my letter and pay nothing. I responded in accordance with the disputes clause in the contract with a three-day notice that the project would stop as a result of no payment. The notice was delivered on Thursday, which meant that the project would not proceed the following Monday. That brought everything to a head and the owner called for a meeting on Saturday morning. I responded that I did not normally work Saturdays (which was a fib), but that I would be happy to oblige.

As the meeting began, the chairman of the owner's building committee expressed disappointment about the work stoppage. One of the committee members added, and I quote: *If we knew you needed the money that badly we would never have given you the job.* I remained calm and responded: *This meeting has nothing to do with a need for money. This meeting is about your breach of contract.* Voices were raised in frustration and exasperation. When they came back to order the chairman said that while it was actually none of my business, they would explain why they could not make payments within the contract time frame. The reason was that their building funds were tied up in investments that were not available on short notice.

The withdrawal of funds required a minimum of 30-days' notice and could only be withdrawn on the first day of any month. They went on to explain that after they received an approved payment requisition from the architect, they could request payment from the investment firm and receive payment on the first day of the next full month. They would then deposit the check in their building account and when it cleared in five to ten days they could pay me. So they said that the earliest they could pay would be 65 or 70 days after the month invoiced for. I could not resist beginning my response explaining that I had not said anything about the building committee not having the money because as they said *it is none of my business*. All I said was that circumstance is caused by the owner, this committee being in breach of contract. Voices were raised again.

When calm was restored, the chairman asked the architect about the discrepancy (redlining) of the amounts being invoiced for and he responded, My site visit was on the sixth day of the month so *I'm not sure exactly what was completed by the last day of last month, so I had to reduce the amounts to protect you (the owner)*. I was asked what I had to say about that, and I responded that I have an opinion about what was completed on the sixth day of the month. I conducted a detailed inspection of the site at the end of the last day of the month, as prescribed in your contract, and carefully measured, counted, and photographed the work in order to produce an exact invoice. That ended the discussion about the amount due, presumably because the results were obvious to all.

The owners' committee had a private discussion among themselves and then reported that they had enough funds available for this rather small first invoice and that I would be paid sometime next week. As some got up from their chairs to leave, one committee member stopped the others by saying loudly to their chairman; As*k him if they are going to go back to work on Monday.* The chairman said, *I think we settled that*. The member repeated, *ask him!* I was asked and very calmly responded, *No*. Voices rose again and then the chair asked *why not!* My response was: *I don't have an active contract because the contract is breached. And before you ask, we will go back to work when we are paid in accordance with the contract that you wrote.*

It turns out that the local funds that they had were available, but they were not sure whose signatures they needed. They decided among themselves that no matter whose signatures were required, they could have the check signed and hand delivered by 10 AM Monday morning. The chair turned to me and politely asked if that was satisfactory and would the job start again first thing Monday morning? I said certainly, I'm a reasonable man. We were paid accordingly and paid before the due date every month for the rest of the project. The chairman then asked the designer: *How can we know in advance how much to have available each month?* The designer answered that he didn't know, but I responded that I could provide them in advance with the approximate amount for every month of the project. The result: We were paid before the due date for the life of the project and the designer never again redlined an invoice. Was the inspection on the job any more difficult than usual? I'm not sure. However, what I am sure of is that construction contractors are entitled to be paid in accordance with their contract.

Slow Pay

Slow pay and retainage being held too long are an increasing problem in the construction industry. And part of the problem is that too many contractors are allowing slow pay to continue. What I hear constantly is: *We can't do anything about it.* Industries-wide, few get paid on time because construction professionals do not enforce the payment clauses in their contracts. Payment clauses are one of few contract-clauses that are in favor of the contractor. If we do not enforce our contract rights, we can actually lose those rights. A case in point.

Some years ago, a contractor remained unpaid for the first four months of a project. After threatening for months to pull off the job, a payment was finally received in the fifth month. However, it was for the first month's invoice only. Continuing to complain, a month later a second payment was received, for the second invoice. This continued into the 10th month of the project until the contractor could no longer borrow additional money to fund the project and finally pulled off the job. The project owner hired another contractor and completed the project. The original contractor sued and more than three and a half years later the court ruled that the contractor did not have the right to abandon the project. The court ruled that under legal principles the contract had been altered. The contractor, by his actions of accepting payment five months late for 10 months had altered the contract which the court said now read; *invoice by the first of the month to be paid five months later.* While readers may say the ruling makes no sense; it was based on contract law and was not appealed.

Each construction professional must decide for themself what their approach to this subject should be. There is a direct cost to contractors resulting from slow pay. There is the cost of interest paid on lines of credit required because progress payments are overdue or long overdue retainage. The cost is astronomical. Think of the interest that could be saved if contractors were paid on time. And the interest that could be earned if the funds were in the contractors' accounts instead of the project owners. The amounts would be staggering, and most construction businesses would be out of debt.

One of the first things construction professionals can do to collect project payments faster is to prepare and send out requisitions for payment on time. Research indicates that industry invoices regularly go out on the 5th and 6th, even the10th of the month or later. This when contracts specifically state: *Bill for work completed as of the last day of the month, to be paid no later than the 15th, 20th, 25th, or whatever of the following month.* When I point this out to contractors, I often hear something like; *what's the difference, owners aren't going to pay on time anyway.* We can't expect designers or owners to begin the payment process until they receive the payment application.

How to Get Paid on Time

You need to begin the billing preparation well before the last day of each month. Have a tentative requisition done by the 26th or 27th, anticipating work that will be completed by the last day. Verify later by phone that it is completed. In that way the requisition can be sent by fastest method on the last day of the month or first day of the next month. Hand delivered if practical. Call the designer to make sure the invoice is received and isn't sitting at someone's desk. Getting results depends on letting people know that you fully expect to be paid in accordance with the contract. Designers and owners understand that timely invoices are to be expected from an efficient businessperson. I don't recommend starting this process on existing projects because it is difficult to change attitudes in mid-job after everyone's gotten used to a way of doing things. Your clients are already in bad habits that you have allowed so it is critical to start from the first requisition on the next job. And then be consistent in expecting timely payment throughout the job.

Figure 5.2 Timely payment is critical to contractor cash flow and survival, yet too often delayed or reduced.

There is nothing wrong with expecting to be paid in accordance with your contract and asking, even demanding, payment (Figure 5.2). Announce your expectations at the initial job meeting. After everyone tells you what they expect, tell them what you expect. Be up-front and say, *I want to talk about the payment process*. Your payment is not going to be important to anyone unless you make it important. Unless you let everyone know you're not embarrassed about being paid on time, it will never be discussed until payments are overdue. By then you have lost the initiative and late payment is the norm, because you have allowed it to become the norm. Payment is not a backroom discussion. Explain that your efficiency and productivity depend on paying your subs and suppliers on time and that you have no intention of being an investor for this project. You have only been contracted to build it. The owners and designers should know this and shouldn't need to be told. But they do have to be told. If you don't manage the payment process and keep it in the forefront of everyone's mind, no one is going to do it for you.

More Payment Case Studies

I was the successful bidder on a large community college building and the designer was one of the largest in the state with a reputation for being tough on contractors. Toward the end of the first on-site job meeting, I said I wanted to discuss the payment process. The designer said you haven't built anything yet and you want to be paid. I responded that I expected to be paid on time in accordance with the contract. As he got out of his chair he said: *Don't be ridiculous, the state doesn't pay anyone on time.* Then he turned and said as sarcastically as possible while leaving the trailer: *Take it up with the attorney general*. It was clear he did not mean that literally.

I took him up on his comment and wrote to the Attorney General outlining my position, enclosing a copy of the payment provision in the contract and quoting the designer saying the state doesn't pay on time. To add weight to the importance of communication I sent it via certified mail, return receipt requested. By coincidence and luckily, I had a friend on staff at the Attorney General's office, so I heard what transpired at the next weekly staff meeting. A young staff attorney reviewed my letter and reported it at the meeting. The Attorney General said *I assume the state pays in accordance with its contracts so advise whoever is responsible for that project to pay in accordance with the contract.*

I never heard anything further from the designer about payment, but we and all the other prime contractors on the job were paid on time every month for the 20 months the project was scheduled for. What is interesting is that several of the prime contractors had *work-in-progress* on other state projects and all of them said that this was the only project where they were being paid on time. The lesson is that you must assert your contractual rights because they are not necessarily automatic. This may partially answer why so many contractors are not paid in accordance with their contracts.

On another public project, I asked a public official about a project payment which was about to come due. He responded that the payment was approved but that he could not release it because it was not due yet. I produced the contract and showed him the payment section which read, *pay by the 20th of each month.* He said: *Then what is the due date*? I replied that the contract clearly states payment is due upon approval, but no later than the 20th of the month. Without another word he handed me the check. I'm not suggesting that getting paid is easy by any means, but I am suggesting that we need to be aggressive in collecting the money we are entitled to by contract. I am convinced that because contractors' industry wide are paid late so regularly we actually *expect to be paid late, and worse, accept it as business-as-usual.* It is our *responsibility* to read and understand our contracts, particularly the payment provisions, and if we expect compliance, it is almost a necessity to explain some clauses to the owners.

How to Get Paid on Time and the Cost of Late Payments

It is necessary to clarify the contract payment procedures to your owners before the first requisition is due and clarify your needs and expectations. It is advisable to be exact on first invoice and even low, particularly if there are any partially completed items that may not be acceptable yet. That means a requisition on time for exact quantities or less of the work completed. This way, if there are issues with the designer, you can move with confidence right from the start, knowing they will be proven wrong if the requisition is red lined. They'll get accustomed to taking your word for quantities of work completed. You'll get a reputation for fair rather than inflated requisitions. Almost all payment clauses say that payment is to be made by the 15th, 20th, or such, but many owners take the attitude that it is not really due on that date because they assume that it isn't overdue until some kind of *grace period,* like ten days after the date. In spite of owners who feel that way, *by* the 20th means just that. And it will continue to mean just that in the construction industry only if we insist on it.

The cost of accepting late payments should be obvious. However, there is also a cost for the delayed release of retainage that often is greater than the amount of interest the money could have earned. When retainage delays get stretched out and the need for the funds increases, the value of the retainage begins to outweigh the value of inappropriate punch list items. By that I mean that a lot of contractors become desperate enough to get paid that they do questionable punch list work they don't really owe the owner or cut a deal forgiving some retainage in return for collecting the rest. The forgiven amounts are ostensibly concessions, but too often *give aways.* This happens even on jobs where a contractor cooperates throughout the project attempting to *appease* designers or inspectors. This is why I so strongly recommend standing firm for your rights throughout the project and particularly on payment requirements from the first requisition. Not getting paid in a timely manner creates cash flow issues, increases project risk and takes the fun out of the construction business.

Cash flow is always a concern, but when a contractor's marketplace weakens and their work slows down, cash flow problems can become critical very quickly. If all of a contractor's reserves are tied up in receivables, it can force you out of business. If some receivables are old retainage and slow payments, the contractor shoulders part of the blame for not *demanding* what is rightfully theirs. When an owner doesn't fulfill their end of a contract, a completed project is excellent collateral. It makes no sense to struggle financially because someone is failing to honor a contract that you've completed.

Know Your Rights

I regularly meet construction professionals who tell me that they do not read the contract documents for each project. I should be shocked but I am not. Some say things like: *The contract is full of legal nonsense*, or *I can't understand it*, or worse, *I have my lawyer read it*. If you don't read the document that legally binds your firm's performance, how do you know what is expected, inferred, or implied? Another response I hear is that *contracts are all the same.* Which is simply not so. The excuse *there is nothing I can do about it anyway* doesn't cut it. The fact that you can't change it does not relieve the obligation to live up to it. And you cannot live up to it if you have not read it.

Reading your contract is a critical *defensive action.* Simple example: A common clause I find, states in one form or another: *The contactor shall review the plans, specifications, and contract and advise the designer if there are any conflicts. Failure to do so precludes using conflicts as a defense later.* After I read a contract, I always send a letter or email to the designer. In an attempt to respond and potentially nullify the above clause I write to the designer something like: After referencing the clause stating, *I have studied the plans, specifications, and contract and was unable to uncover any conflicts. However, I am not a licensed designer and therefore not qualified to discover conflicts. As the designer and author of these documents please advise me of any conflicts you know exist. If I do not hear back from you, I will rely on that as an assurance that no conflicts exist.*

As you can see this demonstrates compliance with the clause and places the responsibility (or attempts to) back where it belongs. The clause as it appears is not easy to enforce, but with this response it is very close to impossible to enforce. I often hear from construction professionals that they don't have the time to do all of the defensive actions I recommend. My response is: In the second highest failure rate industry in the nation, you don't have the option to skip defensive actions.

Summary

The construction industry cannot thrive on delayed compensation or the silent acceptance of contract breaches. Contractors must understand their rights. They need to interpret their contracts and take a firm and consistent stand from day one to ensure timely payment for their work. Being "professional" doesn't mean being passive. Being professional means requiring all parties to the contract to perform in accordance with the terms of the contract. This includes being assertive in upholding payment clauses. By refusing to normalize late payments or the arbitrary red lining of payment requisitions, contractors not only protect their bottom line but help set new expectations for accountability and fairness in the industry.

6 Accuracy of Performance Data

Introduction

Too many contractors believe they have to generate formal annual financial statements for the sole purpose of satisfying the requirements of their bank and bonding company. So, they have their accounting firm produce the statements and send them off. The contractor then has lunch a few weeks or months later to talk about how things are going with the business. A few have said that they learn over time that there are some things in the financial statements that seem to catch the attention of their credit grantor, such as old receivables, declining sales, payables exceeding receivables, etc. Over time some contractors claim to have "learned" to manipulate numbers that *make those external stakeholders happy*, which seems a bit skeptical. A few have actually said that a contractor's financial statements *need to be optimistic* because banks or bonding companies may start constricting their credit limits if the statements are less than optimistic. (*Note that financial statements produced by independent outside accounting firms are considered to be accurate and never subject to optimism, pessimism or suppositions of any kind.*)

The primary purpose of annual financial statements is a management tool for contractors and their managers to use in planning, decision making, and operating the business. And perhaps more importantly, as addressed later in this chapter prior financial statement from years past provides data and produces a *roadmap* for future success. Few would disagree that the accuracy of the accounting data that contractors' base critical decisions on is paramount. There is no implication here that accounting errors exist. It goes without saying that if any of the input data within the accounting system of a construction company is inaccurate, or if any input data is missing the result will be misleading. The accounting for the information available will be correct but the results will not be correct. Results, extended to two decimal places, tend to encourage confidence and trust in the data and the reported results. This chapter is not about poor accounting practices or processes. This chapter is about the critical importance of connection between accurate and timely accounting information and the success or failure of a construction enterprise. The financial performance of a construction enterprise is measured through accounting data. To correct some industry shortcomings that cause business failures, construction professionals need to:

- Improve quality and accuracy of field information collection
- Accelerate timing of data transferred from field to the office
- Require suppliers to expedite their billing processes
- Demand accurate and timely billing from subcontractors
- Process immediate, accurate invoices for work performed
- Expedite payments by proactive interaction with clients

DOI: 10.1201/9781003500315-7

Managing Accounts Payable

To measure financial performance a contractor must know how much they spend and how much they have earned on a project, which, in construction, is not too difficult a few months after a project is completed. Because by then all of the project invoices have arrived and the contract amount including change orders have been agreed upon. However, to manage a multi-month or multiyear project, it is critical to measure financial performance throughout the life of the project. As mentioned in an earlier chapter; to manage labor costs you need data weekly if not daily. Unfortunately, exact material, subcontractor, miscellaneous and overhead expenses are typically available on a monthly basis. These expenses are invoiced by the vendors monthly and recorded as accounts payable.

From accounts payable the costs are recorded against specific projects in what is commonly referred to as a project *cost control* process. As described in an earlier chapter, the firm's books are closed before monthly totals can be lifted from them and they are closed on a date selected by the firm. This closing date could be described as an *arbitrary* date. If an invoice, small or large, is received after the books are closed, it cannot be included in the prior month and is recorded in the following month. Effectively it is recorded as a cost incurred in the following month so the total costs for the month in question becomes less than the actual costs incurred in performing the work because accounting is missing the costs of these late invoices. Both months are in error with some people arguing that it all works out in the end. This can't be so unless every month the same amount of costs are conveniently received and recorded late. What is indisputable is that the field and the cost control process receive incorrect cost information resulting in a lower cost than actually incurred being reported for last month and a higher cost reported than incurred in the following month.

The lower than actual costs are used in the calculation of unit costs incurred in comparison to the unit cost utilized in the estimate for the work in question. Comparison of estimated to actual unit costs is a simple cost control measure most contractors use to assess how the project is performing. Understating costs incurred to date indicate better performance than estimated, indicating to management the work is progressing well or profitably when that is not the case. Those that argue *it all comes out in the end* would suggest there may be costs in the month in question that were pushed forward that may compensate for the discrepancy. Readers who have been following my approach to matters like this can anticipate my response. That comment only reinforces my evaluation that the process almost anticipates error, which causes one to ask which months, if any, should be trusted?

Add to this the reality that the same type of work does not necessarily continue from month to month, so the costs are for a different activity altogether making their use for any purpose almost comical. Subcontractor and material costs are also considered in cost control. We may want to recall the problem discussed earlier when some vendors do not invoice in some months for tax purposes or because they are waiting for completion of their work before billing. The point I am trying to make is that cost control is critical to success and impossible without accuracy and timeliness. I am describing the reality and existing cost control reports are claiming accuracy to two decimal places.

All of the above notwithstanding, there are other exposures to the timing and accuracy on recorded accounts payable. In Chapter 3, we described an invoice approval process used by some construction firms that delays the timely recording of accounts payable in the name of *accuracy*. Readers will recall the example of sending received but not recorded invoices to remote project sites for pre-approval. Because of this process, account payable recordings were delayed, in many cases by weeks and in some cases by months. These were described as examples of accounting errors, but the invoices also distorted the company's cost control processes. This reinforces my position that all cost control must be removed from the general ledger and returned to a field activity. It is the only way to effectively and efficiently manage the work (Figure 6.1).

Figure 6.1 Accurate, timely cost data from the field is essential to effective project control and financial decision making.

Danger of Late Information

When some of the largest construction companies in the country failed, their liabilities exceeded their assets by huge percentages. Be reminded that while these firms were considered solvent, their certified annual financial statements (somehow) verified that their assets exceeded their liabilities. I mentioned this only to point out that these contractors and their accountants believed they were doing fine financially. These failed construction firms relied on their books and records, as any businessperson would, but the books and records were in error. In some cases, by a lot. You can't say that the books and records caused the failure, but they provided confidence and prevented action. What is frightening is that the independent auditors, internal accountants, and contractors were not aware of the losses while they were incurring them. They all relied on the records of the firm and there were no miscalculations in the accounting records. The amounts reported were incomplete because the information was simply received late. As these firms grew the delayed amount of information also grew.

Managing Accounts Receivable

It is a standard procedure in the construction business that invoices for work put in place are recorded as they go out, and the invoices go out on time. Incoming payments are often not received promptly, but accounts receivable records are current. It is fair to say that construction industry accounts receivable are almost always overstated. It is also fair to say that contractors are generally optimists and believe that they will eventually collect all receivables no matter how long they have been on the books. As a result, every failed construction firm I have investigated had large amounts of old (sometimes ancient) uncollectable receivables on their books, receivables that a prudent businessperson would generally have written off long ago.

Adding to the often-overstated accounts receivable issues are invoices that get recorded internally and sent out only to be returned by the designer in a reduced amount. This issue is unique to the construction industry and happens more often than readers may think. Designer-approved invoices are also sometimes paid by the owner of the project in a further reduced amount, for any number of reasons including that the owner did not have the funds to pay the full amount. An accounting or record keeping issue in both of these reduced receivable occurrences is that the majority of contractors do not reduce the amounts recorded in their current accounts receivable ledger. The most common reason given is that *we will collect it next month so way bother*. Some include that because the books for the month in question are already closed their accountants only make the corrections in subsequent months so who cares. The issue here is that this is not an uncommon industry occurrence so too often accounts receivable are overstated—perhaps for a little while or potentially for the life of the project. It is a major issue if our accounting records are ever incorrect, and my experience is the overstatement is not always caught by auditors. I had one auditor say to me, so what, it's only a small amount. My reaction to that is that these records are supposed to be accurate to two decimal places. No wonder I'm skeptical.

Having reviewed tens of thousands of construction business financial statements, I can say without reservation that construction industry accounts receivable are very close to always overstated. And for struggling firms, they are always overstated and sometimes by a lot. To reinforce this position, in my entire career I looked for and never found one example of understated receivables. Obviously overstating receivables overstates profits for the accounting period in question. Overstated profits lead to poor project level and business level decisions.

On the other side of the ledger, accounts payable are often recorded late. The primary reason is that some suppliers and a lot of subcontractors, particularly small and mid-size, are inefficient in their billing process. They just don't get their invoices out on time, with many saying *what's the difference, I never get paid on time anyway*. As stated in other chapters there are also tax reasons that subcontractors and suppliers withhold invoices. For small and some mid-size suppliers and subcontractors, delaying invoicing at year end or at their fiscal year end postpones income and therefore any potential tax obligation until the following year. Some readers may think that the contractors know they owe these vendors, but no one can keep track of dozens or even hundreds of vendors. Payables are the concern of the accounting department, and the accountants have little field information to evaluate vendor invoices.

The record keeping issues come down to a matter of balance. If errors in payables and receivables were random, then perhaps the pluses and minuses would average themselves out. Accounting results might then be a close approximation of the correct amounts. However, as demonstrated, accounts receivable are always either correct or overstated. There is no phenomenon that understates receivables. Accounts payable may be correct but are typically understated because of invoices received late or delays in recording them. There is no phenomenon that overstates payables. It is clear that the things that impact payables and receivables are outside of the influence of the accounting department. In fact, they take place out of the sight of the accounting department.

It can be assumed that most accounting department personnel will be surprised by this material and some will challenge it. It is my hope that the examples described may at a minimum cause some construction accountants to take a closer look at how the information they receive is handled from its source to their department. As the anomalies impacting both payables and receivables overstate profits, the impact can be extreme, and these exposures must be clearly associated with construction having the second highest failure rate. The information that reaches the accounting department assures that payables are correct or higher and receivables are correct or lower.

One immediate takeaway for all construction professionals, field and office is to realize that financial statement profit results are best defined as: *The most profit the firm may have earned.* Not the average or even the likely profit, but the most. That automatically causes you to slightly discount financial statement profit which in almost every case with be closer to reality. In an industry with notoriously slim profit margins, this can do no harm. It becomes a motivator. I sincerely hope this does not become controversial. I fear that a movement to defend the status quo is the most likely reaction which will only tend to protect our industry's position as second highest failure rate in the nation.

Corrective Action

The exposure to understated accounts payables cannot be eliminated but it can be controlled and improved. The first step is for both operations and accounting personnel to be made aware of these realities and their input sought about potential solutions. The next step is to collectively agree that rapidly getting invoices recorded, but not necessarily approved, is critical. It is essential to have accurate record keeping, but to accomplish anything, incoming invoices need to be recorded immediately as is. Because payables are understated industry-wide there is little danger that your payables with ever catch up and actually be overstated. And if so—so what. I guarantee they are already wrong now and have only a limited chance to ever be perfect.

One hundred percent of vendors need to be notified that on time accurate invoices are required. Many contractors make it: *A requirement of continuing to be a company vendor.* This obviously and definitely has nothing to do with your company vendor payment practices or policies. This addresses getting and recording the liability, not paying it. However, I will share that construction firms that make an effort to pay on time are generally rewarded with a loyal following of vendors who perform better on the firm's projects than they do on the competition's projects.

Some of the successful firms I have dealt with put incentives into their requirements. Some send out warnings after a late invoice and have taken some who refuse to comply off their approved vendors list. Full compliance is just a matter of making the effort. When you think about it, a business not knowing how much it owes to vendors is almost comical. Most set a deadline for invoices in their purchase orders and subcontracts that state that invoices received after a certain date will not be paid for another month. With operations and accounting working together it is not difficult once you decide to do it. After the initial effort compliance becomes automatic and less effort is required.

Accounting for Overhead Costs

Another cost that too many contractors do not account for correctly, and do not recover completely from projects, is annual overhead. In theory these costs are *incurred monthly* but payment for overhead costs is not paid monthly. Overhead costs are recovered through the fee or markup applied on every project the company establishes a price for. For example, insurance premiums, vehicle expenses and office equipment are company overhead costs incurred and paid for in advance of performing the work, but these costs are recovered through the markup included in the monthly billings to clients and through the receipt of progress payments for work performed. Complicating the allocation of overhead costs to projects is the fact that exact annual overhead expenses are not known at the beginning of the year, or worse, at the time each project is estimated. Additionally, the inability to accurately predict total annual sales volume while bidding new work means you are also unable to accurately determine how much each potential project should carry as its proportional share of the company's total overhead cost. Not knowing in advance what total sales will be for the coming year means that *estimated* overhead costs are applied to *estimated* annual sales.

There is clearly some *guessing* involved in how much overhead to apply to each project. This means serious guessing about the amount of overhead to apply to multiyear projects. Someone may suggest adding additional potential cost to cover the risk, but that increases the chance of pricing too high to capture the work. In my experience few construction organizations pay much attention to these calculations. Total overhead costs are finally known after year-end financial statements are produced and I see few construction professionals with any interest in what is by then considered *history*. So overhead recovery is not a problem until it is a problem. Oddly enough the solutions to accurately calculating appropriate overhead expenses were discovered during an extensive research project about construction industry financial cycles.

Construction Industry Financial Cycles

Since the late 1940s (after WWII) there have been nine Construction Industry market cycles averaging, until recently, 10 plus years of growth market following by one to three years of construction market downturn (negative growth). The current up construction market is the longest one since the 1940s.

I believe the reason why it is not common knowledge that construction has a measurable growth/no growth cycle is because declining markets are so far apart. Ten or twelve years is half the average leadership duration in a construction firm. Anyone with enough experience to enter management seldom experiences two down markets before retirement. Even those experiencing two down markets seem to consider each as unique to the specific economic conditions at the time and sense no pattern.

The reality is the US economy has experienced cycles on average every 10 or 12 years since WWII. Construction is a lag industry which means its market turns down from 12 to 18 months after a US economic slowdown. It also recovers 12 to 18 months after the US economy recovers from a downturn. In the case of a short US economic downturn of less than a year or 18 months, the construction downturn actually occurs after the US economy has recovered. Many construction professionals do not see the very real connection between the two events. The important lesson is that construction market downturns will occur and are predictable. When the US economy turns down, economists often refer to that as an economic *correction*, the construction market will turn down 12 to 18 months later (Figure 6.2). Construction market cycles are described in detail in the book *The Secrets to Construction Business Success*; Schleifer and EL Asmar, 2022, Routledge.

Balance Sheet Values

A financial statement balance sheet lists all of a firm's short- and long-term assets and liabilities that determine the value of a company. While most construction professionals are interested in the income statement results, fewer seem to have concern themselves with the balance sheet information and results. Many seem to presume that as long as a company continues to earn profits, its value will increase. While earnings are certainly a factor, all earnings do not necessarily remain in the corporation. Some earnings may leave the company through distribution to the owner(s) as dividends and some earnings may be converted to non-liquid assets through the purchase of equipment, office furnishings, or real estate. This use of earnings, other than the purchase of equipment, available cash and offers little or no earnings power to the firm.

The least productive asset for a construction enterprise, in the author's opinion, are furnishings sometimes referred to as leasehold improvements. Obviously, some furnishings and office

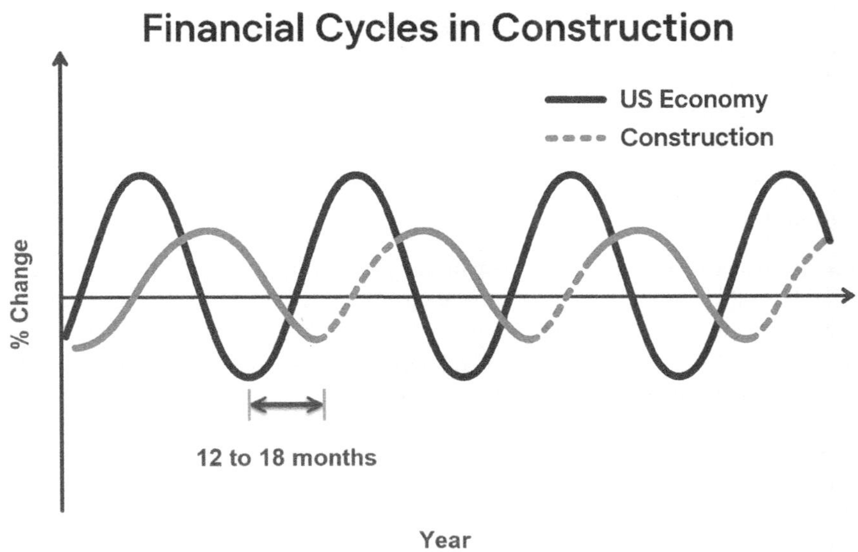

Figure 6.2 Illustration of typical construction industry economic cycles.

equipment are necessary for the operations of the business, but these vary hugely in costs from basic, to comfortable, to elaborate. Almost all startup construction companies are housed in modest rental space with very basic furnishings for obvious reasons, primarily of which is that there is no certainty that the firm will survive. Some begin in the contractor's basement or garage. With modest success space and furnishings are outgrown and advance to the next stage of growth—survival. A majority of firms that survive the more than 50% failure rate of startups graduate to modest rental space expanding their modest office equipment and furnishings.

Construction Business Life Cycles

If a firm graduates from the survival stage, and many do not, they advance to success stage. Which by definition means that for the first time in their existence the firm has positive cash flow. This is because expansion costs money and *growth eats cash*. In the success stage, debt is typically reducing because cash flow is no longer a struggle. The next stage is the growth stage, and it is voluntary. The growth stage is defined as growth in excess of 15% a year because that is the sustainable cutoff point. Growth beyond 15% begins to eat cash again necessitating bowering money to support cash flow. There are lots of reasons for this which can be found in my book *Managing the Profitable Construction Business*, Wiley 2014. After enjoying success stage, some contractors try growing at a rate in excess of 15% per year and find they do not enjoy the stress. They simply reduce size and fall back into success mode. Many do this with no understanding of the *stages of growth* discussed here. It is just intuitive.

Those that advance into growth mode generally do well as they have dealt with the stress and other growth issues before during survival stage. Continuous successful growth is possible, but some growth is incremental. Three years of growth in excess of 15%, followed by reduced or no growth for a year or two, is found to be the easiest. However, market opportunity is usually the primary driver. Growth at this stage requires increase in debt with the cost of the debt covered by the profits from the increasing volume of work. The last stage of growth is *maturity* in which a company is among the largest of their competitors. The company may have debt but is financially stable. Few startup construction companies have ever progressed from startup to maturity in a single generation. Therefore, a mature construction company is the second generation of leadership or older. A complete description of the *construction Company Life Cycle* can be found in my book *Managing the Profitable Construction Business*, Wiley, 2014.

Risk of Losing Focus

An interesting phenomenon in mature companies is that while stable, the majority are seldom the most profitable in their field. It seems that size and stability can lead to a false sense of security and overhead tends to increase for what may be described as *comfort* rather than necessity. Some might describe this shift as a shift from business requirements to luxury. This has the obvious effect of reducing profit margins in these extremely large, privately held organizations. Over time, and sometimes it takes generations, financial struggles ensue as their below industry average profits margins slip slowly (almost unnoticeably) and as their comfort-driven overhead increases. To the average industry professional this may seem unlikely. However, as someone involved with financially or failing construction firms, I can assure you it happens more often than you might guess.

After generations of hard-earned success, firms finally have time to reap some of the benefits secured. Some, but certainly not all, of the most successful closely held businesses opt for a comfortable corporate *lifestyle* that may include things like non-working stockholders and family members on payroll, luxury company cars, corporate aircraft, etc. (Figure 6.3). An example may

Figure 6.3 Excessive overhead and comfort costs can erode profitability, even in otherwise successful construction firms.

best define what I am talking about. This substantial construction organization occupied attractive offices in prestigious buildings. All of the senior vice presidents had a private office with adjoining private bathrooms with a shower. The vice presidents' private offices had private baths without shower, and the assistant vice presidents' private offices shared a private bath with an assistant vice president in the next office. Individual offices were exceptionally large, and furnishings throughout could only be described as spectacular. Company vehicles started at the top with senior executives having imported luxury cars and at the bottom with domestic luxury cars. And their corporate aircraft were all stored in a huge private hanger at the local airport.

My purpose in describing this firm is not to be judgmental. However, I should add that I was visiting on behalf of a client, which could only mean that this firm might be experiencing some financial stress. This *comfort* cost issue is not restricted to mature construction businesses. It has also been found in companies who remain in the success stage without moving into the growth stage, or that tried the growth stage and found the stress not worth the effort. A company spending a length of time in the success phase is, by definition, profitable and is reducing debt or is out of debt. These companies would accumulate cash, and sometimes elect to spend it on comforts such as more luxurious office space, furniture, and/or equipment. The one-time costs can of course be depreciated over time, but an honest definition of comfort costs is *unnecessary expenses.* These non-productive expensive are allowable under tax regulations and are depreciated over time but being unnecessary can also be described as waste. They consume and asset (cash) and add comfort, which is not a tangible asset therefore it is considered waste or loss. The expense is only harmful if the contractor does not understand that in a strict definition it is *waste.*

I am the first to admit that the reward for hard work leading to success should include some comfort. My point in this chapter is that it is critical that construction professionals know exactly where they stand financially in order to measure their performance and manage their spending. And that is impossible if even a portion of what I am describing in this chapter is happening in your company. If any information is late getting into an excellently managed accounting system, all of the accounting reports are in error. You can benefit from knowing how erroneous or late information has contaminated the results of literally thousands of construction firm financial statements. This is not speculation or a discussion of possibilities. These are facts taken from certified financial statements that I or experts reporting to me have personally analyzed.

Every reader should ask their accountant about these issues. The usual response is an explanation about how accounting is it is an exacting science. I agree with those who say that but usually ask them what the expression *Garbage in—Garbage out* means. I am not accusing accounting systems, or inside accounting staff, or independent outside accounting auditors of being at fault. I am placing the blame where it belongs—on us. The blame lies with us construction professionals who provide our internal accounting departments with the input data they base their accounting on. Many readers may believe that putting the work in place for a profit is a construction professional's only function and responsibility. Consider this: If you must know if your work put in place earned a profit and how much, then should providing accurate and timely project performance data also be included in your primary function? This is not a trick question. How else would your firm's accounting department get the information? Is the accounting department represented in the field?

You can theoretically put the work in place without financial data. But you can't put the work in place for a profit without financial data. Accounting cannot provide you with the profit amount without you providing them with input data. Therefore, the statement: *Put the work in place for a profit* is meaningless. To have a useful purpose it must be: *Put the work in place and provide field input data to accounting to determine if a profit was earned.* This is not a word game. It is an attempt to clarify the critical part field data plays in accounting. Because accounting is an exact science, we trust it. But in construction we seem to have completely missed the reality that the total responsibility for accurate accounting begins with timely and accurate initial field data collection. Much of which is initiated with a scrap of paper torn from a lunch bag, written with a dull pencil, while standing in a muddy trench. I exaggerate, but you get the point.

Conclusion—How Did We Get Here

I fully realize that much of this text sounds like complaints. It is not. I have been trying to answer the question for 50 years. How can the oldest industry known to mankind (construction), that produces some of the most sophisticated structures in the world, have the second highest failure rate in the nation? It makes no sense. I wrote a book about the major causes of construction business failures in 1987 and several since clarifying major and minor causes. The research uncovered numerous industry practices and activities that could be described as minor causes that contributed to the problem. Many were self-inflicted and others detracted from the image of our industry. So, for varying reasons, not the least of which was a desire not to offend the industry practitioners that I respect so much, I did not expound on what might be considered offensive. Many were not easy to explain as anyone who has stayed with me to this point can attest to. I have of late decided that these industry shortcomings should be shared with an ever more sophisticated construction audience.

The construction industry differs from all other major industries in the nation. Construction is made up of over 1.3 million separate businesses where other major industries like airline manufacturing comprises only a handful of firms. Many other industries are limited to hundreds of firms, but construction is an extremely horizontal industry with some big players and the majority of companies being small and mid-size firms. There is a large body of knowledge about the design and engineering of construction but much less about the production of the work. There is even less about the business side of a construction business. Further complicating the industry is that successful contractors and construction professionals are educated and skilled at producing the work. These professionals come up through their specific trade, graduate from a university construction or engineering program. None of these offer any in-depth business, economics, finance or accounting courses. To be successful at putting the work in place a construction professional has little or no time for a solid business education. Many will tell you that those subjects are intuitive, or they figure it out as they go. Interesting thoughts but they are defensive and wrong. I have asked successful construction professionals for decades what debits and credits mean in accounting and have not found ten individuals who knew.

The reality is you can put construction work in place successfully without business knowledge. But you can't run a business of any kind without business knowledge, which is learned not intuited. Obviously, readers don't have to believe this but if you don't believe it, I doubt this book will be of much value to you. Believing you know how to run a business successfully because you are running a business does not make it so. What it does is prevent you from finding out what you don't know. The strongest contractors I have known have a technical undergraduate education and then an MBA (Master's in Business Administration). The second strongest contractors I have known knew they lacked business education and sought out and listened to people who did. This may well be one of the most important lessons in this book. Not knowing something that you need is problematic. However, not knowing that you don't know something that you need is worse; because you think you do know it or can figure it out.

The industry's shortage of business knowledge, let's call it a shortage of advanced business knowledge, is easily fixed. As I have advocated throughout this book small and midsize contractors need to stop acting alone and form small internal discussion groups. By this I do not mean seeking out advice and then making decisions. I mean learning to facilitate group discussion to the extent that the group can assist with and potentially make decisions. If you don't trust your key people, you need different key people. If you are convinced you are the only one qualified to make the decision, research and history would disagree. Will every group decision be correct? No. Was every decision you ever made, correct? Probably not. This does not mean giving up control. You are still the boss and can still overrule anything. What it means is that you will build and strengthen your team, grow the business with confidence and be able to get more sleep at night.

7 Prior Financial Statements

Active Management Tool

Introduction

Constructors are some of the most creative people in the world, but for some reason we were never taught the significance and importance of our annual financial statements. Few are skilled in interpreting the critical performance information encoded in them; and only a handful, if any of us, were ever taught that is critical to the effective management of a construction enterprise. Contractors know enough advanced accounting to decode the huge amount of data presented. Many contractors seem to think that financial statements are nothing more than *report cards* or worse, useless historical data. I have heard many times; *by the time I get the annual statement it is old news.*

Financial statements are the most underused tool contractors and senior managers have at their disposal. Do I need to repeat that? The theme of this book has been that we construction professionals' pride ourselves on being builders, and rightfully so. We are part of the industry that produced the built environment from the beginning of the world. A proud heritage for an industry that has been and continues to hold the record as having the second highest failure rate of any industry in the country—second only to restaurants. A big part of the problem is that we are builders not businesspeople. Our energy is used to produce the work at a profit, not to manage a business. I hear this comment enough times that it should probably be our motto: *If we can build the work, the business will take care of itself.* Catchy but not true.

In Chapter 6, we discussed how any casual handling of performance input data distorts reported financial results. We also demonstrated that if (or when) erroneous input data enters our accounting systems it *always* exaggerates profits and reports better performance than actually occurred. Our industry is notorious for not utilizing the huge amount of financial data we collect. We have elaborate data collection and reporting systems and spend countless hours gathering and processing information. At the same time most construction professionals complain that the information they receive arrives too long after the work is put in place to be of any value to them.

Financial Statements

As discussed earlier, accounts payable are received, stored and reported on by our accounting departments, who also produce, or in some cases just record, the invoices for the work we put in place. The invoices become our accounts receivable. All money spent by or received by a construction enterprise is recorded in the accounting department. From the accumulated accounting data, the company's financial statements are produced. There are three statements: the income statement, the balance sheet, and the cash flow statement (which some of us will remember used to be referred to as the sources and uses of funds statement). For readers who are unfamiliar with them, the following is a very basic description of typical financial statements.

DOI: 10.1201/9781003500315-8

There are three primary financial statements which indicate the financial condition of a company under Generally Accepted Accounting Principles (GAAP), also referred to as accounting rules. The income statement shows the company's profitability for a given period. The balance sheet shows a company's assets, liabilities, and shareholders' equity at a particular point in time. The cash flow statement shows cash movements from operating, investing, and financing activities.

Income Statement

The income statement, also known as the profit and loss statement, is a financial report that summarizes a company's revenues and expenses over a specific period of time—usually monthly, quarterly, or annually. It shows how much money the company brought in (revenue), how much it spent to generate that revenue (cost of goods sold and operating expenses), and what's left over as profit or loss (net income). In essence, it tells the story of whether the company made money or lost money during the reporting period. For contractors, it's a crucial snapshot of profitability and can reveal trends in margins and overhead that influence long-term success.

- *Revenue*: Total amount earned from the sale of goods or services
- *Cost of goods sold*: Direct cost of producing goods or services sold
- *Gross profit (GP)*: Subtracting cost of goods sold from revenue
- *Operating expenses*: Costs of operating the business
- *Operating income*: Earnings, less operating expenses, interest, and taxes
- *Net profit*: Profit earned from revenue less expenses
- *General and administrative (G&A) expenses*: Costs of running the business
- *Pre-tax income*: Earnings before taxes
- *Taxes*: Amount paid in taxes

Balance Sheet

The Balance Sheet displays a company's assets, liabilities, and shareholder's equity at a point in time. The balance sheet gives an overview of what a company owns and what they owe to others. In order to be *balanced*, a company's assets must equal liabilities and equity together.

- *Assets*: Any resource a company owns that has economic value, including what the business is owed such as accounts receivable balances.
- *Liabilities*: Obligations or claims to a company's assets, like mortgages, credit card debt, accrued expenses, payroll, accounts payable.
- *Shareholders' equity*: Amount paid to shareholders if all assets were liquidated, and liabilities were paid.

Cash Flow Statement

The Cash Flow statement displays increases from regular business activities and decreases in cash from spending, changes in cash per period, and starting and ending cash balance. It presents the amount of cash generated through operations, investing activities, and financing activities called net cash flow.

Annual financial statements contain a huge amount of financial and performance information which, in the construction industry, is seriously underutilized. The biggest reason for this is that in our industry, financial statements have somehow become closely guarded secrets. In most firms they are not shared with anyone beyond company owners, and if shared further, only with a very few senior executives. In other industries senior executive and thought leaders thrive on performance feedback because everyone seems to need (and wants) to know how they are doing. That's why there are report cards in education. The reason for not sharing financial information with employees in the construction industry seems to be *it is none of their business*. And a commonly heard excuse is, that the information is received too late to be of any value.

Non-Use of Financial Statements

The typical annual audited financial statement is not issued for 90 days or more after the end of the fiscal year, and unaudited monthly financial reports are seldom issued before the 20th of the following month. We have already discussed the need for project cost control to be moved out of the accounting department because the 20th of the following month is too late. So, let's concentrate on the annual financial statement. My experience is that non-accounting executives have limited interest in the annual statements, claiming that the information is too old to be of use to them. This may be less important because so few contractors or construction executives are able to effectively read or interpret financial statements. I can hear the objections already, o*f course we know how to read them*. I mentioned earlier in this text that field tests among construction executives indicate limited, if any, understanding of the meanings of *debits* or *credits*. There is no evidence that GAAP rule prepared financial data can be accurately interpreted by anyone who is not trained in accounting, economics, and finance.

In specific construction industry field tests, many (but not all) contractors and construction professionals were able to understand the basics of an *income statement*, few could correctly define the use of the *balance sheet*, and the majority did not know that there was a *cash flow statement* included in the financial statements of their company. This was observed among contractors who claim that they fully understand financial statements. When this subject is pursued further with members of the same group, it usually results in the admission of very limited use of the type of information contained in annual financial statements with the exception of gross and net income and G&A expenses. The general consensus expressed was: *The only thing we* (construction professionals) *care about or need to know is how the projects are doing. There is really no practical or useful information on the balance sheet*. And most simply admitted that they had no idea what was supposed to be in the cash flow statement.

It was also learned that most non-owner construction professionals never saw their firm's financial statements. The primary reason expressed beyond its none of their business was some form of: *If I shared my financial statements,* s*uppose they were to get into the hands of my competitors?* Some added: *They would know all about us*. Let's pause for a moment and consider this. First of all, the full financial statements of all publicly traded companies are available to everyone. Which is considered harmless. If the fear is that competitors will learn what profit was earned, the question becomes: *What exactly would a competitor be able to do with that knowledge*? I supposed we could speculate. If your profit margin is higher than your competitor's, then their reaction might be jealousy. If your profit is less than their profit, I suppose they would be pleased. It is hard to imagine what use that information could possibly be for a competitor. Or to anyone else for that matter. The people most interested in the performance results and who can most benefit from that knowledge are our employees. Which of course would benefit your company.

I have been told by contractors that *I don't want my employees to know how much I make*. I respond with data from a research project I did some years ago that revealed that construction industry employees believed that contractors make on average 25% profit or more. Some speculated as high as 50%. It would clearly be to every contractor's advantage for employees to learn that industry profits are nowhere near that high. I have convinced thousands of contractors to share their financial information among their employees, and all have reported increases in production and improvements in morale, commitment, and retention. There has been no negative feedback. It is a fundamental motivation principle that people desire feedback on their performance. Performance evaluation has long been accepted as the foundation of performance and process improvement.

Use of Financial Statements

The first step in the effective use of financial information is to understand that it has no value unless it is shared. If you still insist it must be kept secret, you may want to skip this section. The second step is also controversial, which is that a single financial statement has very limited usefulness. The three statements complement each other. However, without some formal accounting education, it is difficult to extract actionable information from the statement of cash flows. Therefore, I will not be spending any time on that statement here.

A Single Financial Statement Has Very Limited Usefulness

In my experience, shortly after an annual financial statement is available, contractors and senior executives may spend some time reviewing often just the income statement. I do not mean analyzing or even studying it. By review I mean casually perusing it and by it, I mean only the income statement. Those going straight to the income statement are usually trying to learn what was earned (profit) for the year (or period). A reader of the income statement obviously has no choice but to accept the amounts that appear in the statement. No matter what the reviewer's personal knowledge about the projects or the firm, the audited financial statement is the final word on the financial performance.

Therefore, they are not *checking* or *evaluating* the statement. They are *reading* what it reports. Reviewing is a passive activity which tends to minimize the need to even look at the statement. This is probably why in many cases annual financial statements are simply filed. Some are not even circulated to management who receive a summary. To be on file seems to be the statement's only purpose other than to be delivered to the surety company and bank who require statements as a prerequisite for doing business. I do not criticize this practice because a single financial statement has very limited value. I cannot analyze a construction firm without three years of financial statements, and I try not to attempt an analysis without five years (Figure 7.1).

I have found a very limited number of construction accounting or finance experts in my career that know how to utilize a running analysis of five-years of financial statements. Why? Because that has meaning. It offers actionable information. I would go so far as to say that only the analysis of five years of financial statements has any value in discovering meaningful, actionable information. When I can only get three years of statements, I use them. But two or one year afford very limited information. I continue to be shocked that few construction professionals have never even heard of this analytical process. Those of us who do engage in this process can't wait to add the next current statement data to our analysis.

The Analysis of Five Consecutive Years of Financial Statements has Great Value and Offers Meaningful and Actionable Information.

Financial Statement Analysis

There are a number of financial ratios that industry experts favor to measure a construction company's performance, the primary being the percentage of profit, both gross and net. One use is to compare a firm's ratios to similar firm's ratios or to industry averages. Most trade associations collect anonymous annual data from their membership and publish the averages including highs and

Figure 7.1 Multi-year financial statement analysis provides meaningful insight into trends, strengths, and risks.

lows. These are good comparisons with one caution. They tend to be a little high in that contractor's self-report, and in weak years or losing years they tend not to report. The averages reported are accurate, but if there are any unreported lower performances they would have reduced profits slightly. All concerned deny this, but it tends to skew the average to the higher profit side. Hoping not to offend anyone, I just lower the reported averages slightly in my mind when I compare a firm's data to it.

An unfortunate weakness in data collection is that contractors manage to have the freedom to record their costs and expenses in more than one method that can vary slightly. For example, some contractors charge project manager's time or company truck expenses to each project while others charge the same expenses to company overhead. Some companies will charge payroll preparation to overhead and others apply it as a project cost. The list is endless, and while no method is wrong, there is no way to separate or even know which methods any firm is using. It all comes out the same in the end, but certain categories of costs may vary. For that reason, I use the nationwide numbers as a cross check or as added information, but not as a primary test of a firm's performance results.

Recommended Performance Test

The performance evaluation process I use and recommend tests an organization's performance against itself. You may want to consider the process before passing judgment. I include five years of financial statement data to calculate all the ratios that I intend to review. For example, to test an organization's profit performance I will list gross profit as a percentage of sales, net profit as a percentage of sales and overhead as a percentage of sales for the four prior years and the current year just added. I have no expectation that the amounts or percentages will, or even should be, the same or similar each year. I am not comparing the amounts or percentages against another group. I am testing the company's performance against itself to determine the answer to the critical question, is the organization's performance improving over time, staying the same, or deteriorating over time?

I fill in the amounts of sales, gross profit, general and administrative (G&A) expenses, etc., across the top of an excel spread sheet from left to right starting with the oldest year to the current year. The spread sheet will calculate the percentages or other metrics I am interested in. Further to the right the spread sheet will calculate a five-year average of amount and a two or three average depending on my interests. The two-year or three-year decision will be based on any exceptional or interesting changes observed in the year-to-year data. Five years of data from other categories of interest will be listed below these. With this format it is easy to create graphs of any of the five-year entries to visualize the direction and rate of change—increasing, decreasing, or relatively stable over time. I can then create a graph of the five years below the data portion of the spread sheet. If the line goes up and down from year to year within a reasonably small rage it means consistent profit performance. If the line, not necessarily straight, but generally trending up it means improvement in performance over time (over some or all of the five years). If the line is uneven but generally trending down it means that financial performance is deteriorating over time.

I assume that readers will agree that the direction of the financial performance, in this case profit being better, worse or steady, is extremely valuable information (Figure 7.2). Obviously, each year when the chart is undated the graph is extended, and performance trends are defined. Updating new information measures the results of actions taken in previous years and may imply the need for actions to correct any unsatisfactory results.

Standard financial ratios are commonly used to evaluate the operating performance of a business and fall into four primary categories: profitability, asset utilization, liquidity and debt utilization. The ratios that I generally track and find to be most informative are shown below. Your interests may vary depending on your background. In my experience CFOs, outside accountants, contractors or board members have a preference for which financial ratios they like to track. Which ratios are tracked are not as important as the need for consistency. That means continuous tracking of the

	Year 1	Year 2	Year 3	Year 4	Year 5	Year 1	Year 2	Year 3	Year 4	Year 5	Year 1	Year 2	Year 3	Year 4	Year 5
Sales	$8,000	$8,400	$8,200	$8,600	$8,500	$10,000	$10,000	$12,000	$12,000	$12,000	$10,000	$10,000	$8,500	$8,500	$8,500
Gross Profit	$1,600	$1,700	$1,650	$1,750	$1,680	$1,800	$1,800	$2,400	$2,400	$2,400	$2,000	$2,000	$1,800	$1,550	$1,500
G&A Exp	$1,000	$1,050	$1,020	$1,100	$1,040	$1,000	$950	$1,200	$1,100	$1,050	$1,000	$1,050	$1,050	$1,050	$1,050
Net Profit	$600	$650	$630	$650	$640	$800	$850	$1,200	$1,300	$1,350	$1,000	$950	$750	$500	$450
Profit Margin	7.5%	7.7%	7.7%	7.6%	7.5%	8.0%	8.5%	10.0%	10.8%	11.3%	10.0%	9.5%	8.8%	5.9%	5.3%

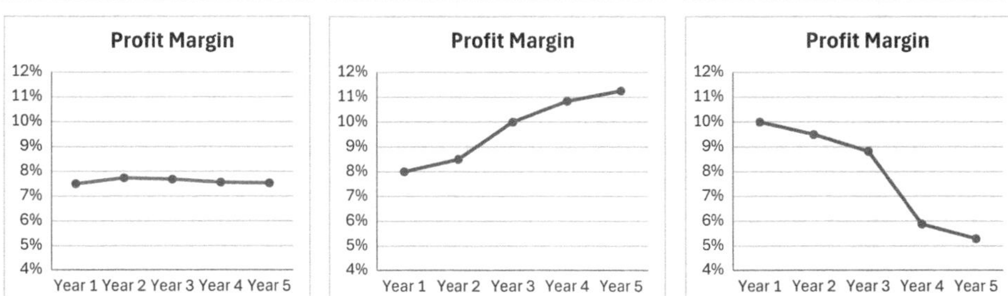

Figure 7.2 Sample graphs showing steady profit margin, profit trending up and profit trending down.

same ratios because they are your preferred measurements of your organization's performance. As already presented, there are weekly and monthly performance indicators, but the ultimate test is the audited financial statements.

For readers who have never used this tool, there is obviously no need to wait five years because your firm's previously audited statements are on file. One cautionary note. Occasionally accountants will change or adjust methods or processes used in statement preparation. If this has occurred within the five years being evaluated, take care to ensure that all numbers used in the evaluation originate from a consistent source or minor adjustments may be necessary. Your internal or external accountants can clarify this for you. Below is a list of ratios commonly used in an evaluation of the financial condition and performance of a construction operation. On the left is what I refer to as the ratio. In the middle is the formula or calculation method to determine the ratio. On the right are guidelines for the results I would be looking for. These are *guidelines* only and can vary significantly from company to company. For example, some firms subcontract much of their work while others self-perform almost all their work. A difference like this (and many others) will dramatically alter some ratios.

Financial Ratios Commonly Used in Construction

Gross Profit %	Gross profit ÷ revenue	Higher is better
Net Profit %	Net profit ÷ revenue	Higher is better
Current Ratio	Current assets ÷ current liabilities	Above 1.3 to 1.0
Quick Ratio	Cash + receivables ÷ current liabilities	Above 1.2 to 1.0
Working Capital	Current assets – current liabilities	Higher is better
G&A[1] Percentage	G&A ÷ revenue	Lower is better
Debt to Worth	Total liabilities ÷ stockholders' equity	Below 2.0 to 1.0

[1] G&A is general and administrative expenses

Some Preferred or Caution Guidelines (Rules of Thumb)

Revenue to Equity	Below 12 to 1.0
Revenue to Working Capital	Below 30 to 1.0
Receivables to Payables Commercial GC	Around 1.5 to 1.0
Receivables to Payables Trade or Heavy-Highway	Around 2.0 to 1.0
Receivables to Payables Labor Intensive Trades	Above 3.0 to 1.0

I use all these ratios and guidelines in the analysis of the firms I work with and am therefore familiar with them. Keep in mind that these are industry averages taken from industry sources and like any average, it may not fit every company's practice or preference. For example, I have encountered some construction companies that I considered cash heavy. They had what I considered to be too much cash in the corporation. These firms obviously had no debt and some had cash that had never been in use, even during peak periods. My attitude is that excess cash serves no commercial purpose and is at unnecessary risk being housed in a construction corporation. I also believe that a closely held firm should have some debt included in their capital structure. There are two reasons for this. One, debt reduces the amount of owner's funds at immediate risk. The other reason is that all commercial enterprises need to have some bank credit immediately available for unforeseen emergencies.

Cash heavy companies seldom, if ever, borrow. Absent a borrowing relationship it can take some time to arrange a loan. It is not unheard of for a construction firm to have a sudden need for a cash infusion. I am not at liberty to disclose the number of times this has happened to otherwise solvent contractors who had to turn to their sureties for help. Something that is unfortunately not widely known is that once a surety puts money into a client's firm, the surety can no longer provide bonding. I advise all of my clients to continually maximize their line of credit even when they have little need to borrow. I recommend that they borrow against their line at least once a year (preferably twice) even when they have sufficient cash to cover the obligation. Some consider this unnecessary, and others have referred to the practice as *belt and suspenders,* but for an operating business in the construction industry it is simply prudence.

In a volatile industry like construction, it is simply good business to maintain the largest possible line of credit. An inactive bank relationship is useless to either party. To easiest way to maximize a line of credit is to use it. The borrowing makes it easier to negotiate an increase in the line every year. It is simple to borrow money when you don't need it, but when you suddenly need to borrow, lenders often become cautious and all of a sudden it takes time to arrange a loan. I am certain that some readers will consider this a step too far, but successful businesspeople manage current and potential risks proactively. This process is easy, cheap, and prudent.

Author's Financial Statement Analysis

Internal performance analysis of a construction organization includes, and is dependent upon, an intimate knowledge of the firm's operations, or at a minimum, unlimited access to the operations and those in charge. There is very little known about how pure financial analysis can be of value to contractors and construction professionals. Until now it was a proprietary process that I developed as a consultant to the surety industry. I refer to it as a confidential corporate financial analysis. It was designed to determine the financial strengths and weaknesses of an operating construction firm requiring access to only their annual audited financial statements. Obviously access to the firm's executives and accountants would have been of value; as would access to the construction sites have been of value, but that was not the case.

This method relied exclusively on financial data and included the use of amounts and numbers from both the income statement and balance sheet. Some may assert that these should not be mixed together but the process was approved by a university in the development of a Ph.D. dissertation.

As mentioned earlier, financial evaluation can be complicated by the fact that there are subtle differences in how construction firms record their financial information. This proprietary process was specifically designed to provide accurate results irrespective of the anomalies that may exist in the recorded data. For the first time this evaluation process is available to contractors, construction professionals, and credit grantors. For simplicity the input data is rounded off to three or four significant digits depending on the size of the company. The standard for rounding used is; the numbers five and above are rounded up and the numbers four and below are rounded down. While financial statements are regularly extended to two decimal places, the decimal places are ignored. While no explanation for dropping decimal places should be necessary, it should be noted that the *estimated* cost to complete is used in construction statements to calculate profit. It is our position that an *estimate* cannot be considered accurate to two decimal places. Numbers lifted directly from the audited financial statements are rounded off for use in the analysis.

Below are the categories listed in the left-hand column of the spread sheet. The years are listed in the top row from lowest to current year starting on the left (Figure 7.3).

	Sample Construction Corporation						5-Year	2-Year
	2020	2021	2022	2023	2024	2025	Average	Average
Sales	$7,256,000	$1,349,000	$2,893,000	$5,580,000	$4,996,000	$5,086,000	$3,980,800	$5,041,000
GP	($417,000)	($999,000)	$487,000	$810,000	$290,000	$716,000	$260,800	$503,000
GP %S	-5.7%	-74.1%	16.8%	14.5%	5.8%	14.1%	6.6%	10.0%
G&A	$938,000	$1,086,000	$481,000	$810,000	$554,000	$708,000	$727,800	$631,000
G&A %S	12.9%	80.5%	16.6%	14.5%	11.1%	13.9%	18.3%	12.5%
NP OPS	$153,000	($2,020,000)	$7,000	$15,000	$144,000	$8,000	($369,200)	$76,000
NP OPS % S	2.1%	-149.7%	0.2%	0.3%	2.9%	0.2%	-9.3%	1.5%
Cash+cd's+secur	1,722,000	43,000	27,000	1,418,000	-802,000	1,470,000	$431,200	$334,000
A/R	3,721,000	234,000	271,000	199,000	1,968,000	398,000	$614,000	$1,183,000
A/P	2,753,000	28,000	2,000	1,171,000	688,000	1,526,000	$683,000	$1,107,000
A/R-A/P	$968,000	$206,000	$269,000	($972,000)	$1,280,000	($1,128,000)	($69,000)	$76,000
C+A/R-A/P	$2,690,000	$249,000	$296,000	$446,000	$478,000	$342,000	$362,200	$410,000
CIEB	$0	$0	$0	$0	$0	$0	$0	$0
BIEC	$0	$0	$0	$121,000	$55,000	$0	$35,200	$27,500
BIEC-CIEB	$0	$0	$0	$121,000	$55,000	$0	$35,200	$27,500
Debt	$427,000	$0	$0	$0	$0	$90,000	$18,000	$45,000
C+(A/R-A/P)-D	$2,263,000	$249,000	$296,000	$446,000	$478,000	$252,000	$344,200	$365,000

Figure 7.3 Sample spread sheet with five-year and two-year averages.

Financial evaluation line item descriptions	
S	Sales
GP	Gross Profit
GP %S	Gross Profit Percentage of Sales
G&A	General and Administrative Expenses
G&A %S	General and Administrative Expenses Percentage of Sales
NP OPS	Net Profit (from operations)
NP OPS %S	Net Profit Percentage of Sales
Cash + CE	Cash + Cash Equivalents
A/R	Accounts Receivable
A/P	Accounts Payable
A/R – A/P	Accounts Receivable – Accounts Payable
Cash + A/R – A/P	Cash + Accounts Receivable – Accounts Payable
CIEB	Costs in Excess of Billings
BIEC	Billings in Excess of Costs
BIEC – CIEB	Billings in Excess of Costs – Costs in Excess of Billings
D	Total Debt
C + (A/R – A/P) – D	Cash + (Accounts Receivable – Accounts Payable) – Debt

While filling in the spread sheet it is important to discipline yourself to avoid making any assumptions about the meaning of the numbers until the entire spread sheet is populated. While it can be difficult to avoid making assumptions as you go along, the assumptions tend to bias later analysis. You can of course have someone else enter the data onto the spread sheet as long as their accuracy is trusted. Accuracy is absolutely critical in that any error in any figure will distort results rendering the analysis useless. I personally prefer to place the number into any spread sheets that I will be interpreting. While time consuming, it is the only way I have found in which I trust the input. After populating and double checking the spread sheet, I begin the analysis from the top. As mentioned earlier, the order of the line items is significant in that it is a logical approach discovered by trial and error over decades of evaluating construction companies.

Obviously you know nothing about this firm other than what is in the five prior years of annual financial statements at your disposal. As readers probably know there are numerous notes in an annual statement. The notes are clarifications that the statement preparer thinks the statement reader may need to more fully understand the statements. For example, a note may be a brief explanation about what the company produces or explains what type of construction firm this is. I generally start at the oldest year and work forward reading every page of the entire financial statements document. I have on occasion been sent only the financial information found on the front portion of the statement document. This is not sufficient for an analysis. You must have the entire statements documents including all notes and subsections.

If after studying the oldest year's statement you are comfortable that you have a reasonable understanding of the firm, you may then read the most current statement. If there are no surprises, you can start the financial analysis. However, if you feel the need for clarifications, you may have to read one or all of the other statements until you are comfortable that you understand the makeup of this firm. You can then start the analysis phase which begins with familiarization with the firm. The purpose of the first exploration of the spread sheet is to familiarize yourself with the firm and its financial performance. It is not to *learn* details or come to any conclusion, especially not performance measurement conclusions. This takes some discipline because there is a natural tendency to jump to conclusions about the meaning of the numbers you are seeing. For example, if you start with the first line sales, and sales are dropping off from year to year, you might think the firm is losing money. Then you may want to check that thought by looking ahead at gross and net profit. It serves no purpose at this point because the object is familiarization, so jumping around to satisfy curiosity does not advance the study.

Sales, as mentioned earlier, may be steady, rising, or declining. Your purpose is to determine that trend. Not the cause. Just the trend. Keep in mind you may be dealing with a $20 million a year firm, a $200, $900 million, or larger firm, which at this point does not matter because you are interested in the five-year trend. The next line is gross profit (GP) in dollars which you might skip for now and move on to %GP. The reason for skipping dollars is that both sales and profit fluctuate in dollars, which is of less interest than the percentage of profit. Is it consistent or fluctuating, and in which direction? Next is G&A and again you can skip it and just look at the percentages. Next is cash, which I generally take a quick look at but don't really study because there are so many reasons for cash to go up or down that have little to do with sales or profits and you won't know enough at this stage to learn anything from how cash is trending.

Next is receivables (A/R) and payables (A/P) which generally follow sales up or down but not always so look more closely at A/R – A/P, which also fluctuates for unexplained reasons For example, receivables came in but payables were not processed before the books were closed so you can expect there to not be much correlation. The next line C + A/R – A/P may clarify AR – AP some because if receivables came in but payables were not made then cash would be increased. The sequence of these categories may begin to make sense because as you work through them, they begin to clarify categories above. The next entries of CIEB, BIEC, and BIEC – CIEB may be confusing to readers who are not familiar with where these financial statement entries come from. In brief these are adjustment entries for overbilling or underbilling projects. Contractors for any number of reasons may invoice a project for an amount less than less work produced in a month (underbilling) or for a greater amount (overbilling). There is insufficient space for a more detailed explanation so for those readers who do not understand under and overbilling, I recommend using this as a learning opportunity and seeking answers from someone you work with. Subtracting CIEB from BIEC will indicate the amount net over or under billed.

The next entry, debt is total borrowings or debt outstanding. Because over and underbilling and debt impact cash, the next entry C + (A/R – A/P) – D is intended (as a catch all) to be a measure of short and mid-term liquidity or total cash and cash equivalents. Each construction accounting process or department manages their records in accordance with GAAP rules. However, within those rules some firms may record amounts into distinct categories. This "catch all" as I refer to it, assists in quantifying liquidity.

After the familiarization phase, the analysis begins. You start again from sales and study the entries in any given year in greater detail. I look to see if there are any years with a 15% or greater increase in sales. If so, I expect a serious increase in gross profit. The reason is that when a firm gets busy, it is almost impossible to hire additional overhead people fast enough to keep up with record keeping. What commonly happens is that accounting work slows down from the staff shortage resulting in late posting of accounts payables. Unrecorded costs can make overhead seem to be reduced as a percentage of sales. Receivables which are simpler do not fall behind. Consequently, reported gross profit tends to increase along with net profit. The gross profit increase may be in error because of late recorded costs and reduced G&A percentage may be actual but temporary, until people can be hired.

Clearly there is a tremendous amount that can be learned from a five-year history. If there is exception growth in a given year, looking forward can verify if profits were real or resulted from late reporting of costs. Profits reported in an accounting period that are later found to be overstated create serious problems. To correct accounting errors, it is necessary to post corrective entries in a subsequent accounting period. That means that correcting for over reported profits will show a lower profit in the subsequent period than actually earned. In some cases, a loss will be reported for the subsequent period, when there was actually a profit earned.

The only method allowed to correct errors in an accounting period causes another accounting period to have data posted that is not representative of what actually occurred in that period. This can only be seen in a multi-year analysis of financial statements. In the case here, the drastic reduction in profit in the subsequent period is right in front of you. For management who looked only at each statement as it was issued—one at the time; all they know is that they made a lot of profit in a given year and much less in the next year. They made all of the management decisions during each of those years based on erroneous, inaccurate financial information. In some cases, they were lucky enough to still be in business. In other cases, their businesses failed. But in either case the construction professionals involved still never even knew how poor their financial performance was.

All my examples are true stories. These construction business failures and close calls actually happened. Thousands of them. I hope readers are following this. In the short span of this chapter, we have discovered mistakes made in our industry where second- and third-generation businesses were lost and fortunes squandered. All of the losses were based on erroneous information caused by the contractors and construction professionals and audited by qualified independent accounting firms. Could this be happening to firms all over the country as we speak? These construction professionals did not see it coming. If they had seen it coming, would they have done something about it, preventing catastrophe? My answer is yes.

Contractors know how to dig their way out of a hole. The way to see it coming is simple. We need to bring our financial input reporting 100% up to date and keep it there. It is on us. No one is doing this to us. This is not the fault of our outside accountants. This is not endemic in our industry. The first step is to reverse denial. I am tired of the challenge I hear so often which is: *Are the financial records of every construction company in the nation not accurate?* My answer is: *Of the tens of thousands of construction industry financial statements I have reviewed, 100% of them had amounts (that is plural) entered into their books in the month or months after the work was done or the material was delivered and incorporated into the work.* Some had just a modest amount recorded late so their books were just reasonably inaccurate. Some had a lot of late information, so their books were considerably inaccurate. The more important question is: Why does our industry continue to have the second largest failure rate in the nation?

The construction companies that will fail three, five, or more years from now will have positive, but inaccurate financial statements this year. Obviously, contractors and their top managers are busy and certainly not looking for something else to do and I am certainly not on a campaign to add to their workload. But this book is not just about why construction remains the second highest failure rate industry. It is also about what needs to be done to reverse that. We can't continue to manage our organizations using financial information that may or may not be accurate. Here are some frightening statistics that substantiate my point.

My firm assisted in the investigation of, and often the liquidation of, well over 1,800 plus insolvent construction firms. In this database, all of the firms were successful for a minimum of 10 years in business. A total of 56% were second generation and 11% were third generation or older. We assisted in the liquidation of hundreds upon hundreds of firms under 10 years old, but they are not included in this database. The reason should be obvious. These *startups* never achieved success, so they were never a viable business to begin with. The database consists of almost 2,000 companies that were successful, profitable businesses that failed. What readers need to understand is that every one of these companies had profitable financial statements the year before they failed. And the majority were found to have had inaccurate statements two years before they failed. Some very large firms, three or more years, but they are the exception.

It is almost frightening to hear those financial statements reported a profit that never in fact existed. If these companies had managed their record keeping properly and prevented the late recording of information, their financial losses would have been reported as soon as they occurred instead of a year or two later. Management would have been aware of their poor performance immediately. We know from experience that they would have been able to do something about it. The construction professionals affected would have at least had the opportunity to take corrective action to attempt to save their companies.

We can't go back in time, but we must ask ourselves: Was the cause of almost 2,000 construction business failures poor performance? Or was the cause of these business failures the inefficient handling of financial information? Inefficiency and late data input enabled losses to be reported as profits. Expressed this way it sounds questionable, but it occurred and is fact. Let's use a word picture.

If we create a graph with the contract price of a project (road, bridge, or building) on the vertical access, with the 18-month schedule on the horizontal axis we can chart costs as incurred (Figure 7.4). The cost incurred line will start at the bottom left at zero cost and zero time. As work progresses and costs are incurred the line will proceed to the upper right. Let's say the anticipated

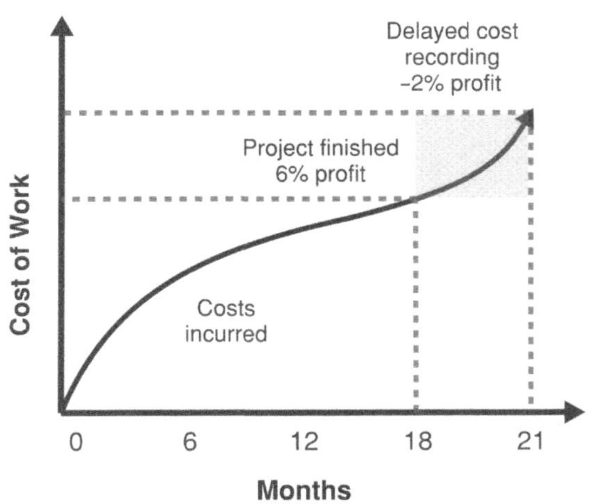

Figure 7.4 Delayed cost recording and the impact on profit.

profit (including home office overhead) is 6%. When the project is completed, if everything goes as planned, the cost line after 18 months will be almost to the top of the graph. Just 6% short of the top when 94% of the contract price has been spent as anticipated.

If the project is completed in 18 months and costs to date at 18 months are recorded at roughly 94% of contract amount, it is time to celebrate a successful project. However, what occurs way too often in our industry is that in month 19, additional costs are recorded for this project. Readers unfamiliar with this scenario are probably not in the senior management of an active construction firm. On too many projects costs continue to be recorded months after a project is finished. In this case costs continued to be recorded in months 20, 21, and beyond. These were not just late material and subcontractor invoices, but additional labor and material costs incurred on punch list items and on work that was not actually 100% completed or accepted by the designer or owner. However, every work line item was listed as finished.

Summary

Construction is a mature industry that takes pride in having produced the built environment that we all enjoy, and the general public along with our labor force and leaders perceive our businesses as prosperous and sophisticated, then it should be no surprise that contractors and construction professionals might be reluctant to even consider any of the industry shortcomings pointed out here. As a lifetime constructor my response is, if our business practices are fine the way they are, why do we continue to have second highest failure rate in the nation? Having devoted much of my industry career searching for the answer to that question, I request your patience. Construction is a great industry, responsible for the built environment, and has been around for a long time. It has perfected the technical side of the industry through centuries of preoccupation with technical challenges. Our industry, of over 1.5 million separate businesses, has not perfected the business side of construction. What we can say is that we can build anything. What we can't say is that we can consistently build everything for a profit.

This book is about the many *business side* issues, as opposed to technical side issues, occurring in our industry which have not been perfected over time. There is not one glaring error. That would be too easy to find and fix. There are numerous less noticeable or more modest business issues that erode or marginalize performance (profit), such as the failure to utilize financial information to measure and maximize performance, and to discover and correct underperformance as addressed in this chapter. These may be perceived as *small* things until something goes wrong because of lack of attention. And they become big things—even business failure.

8 Industry Business Competence Is Weak

Introduction

The backbone of the construction industry is made up of a huge number of distinct, highly specialized trade firms in a subset of specific types of work. A roofing contractor may perform only one type of roof. An electrical contractor may perform only residential, commercial, or transmission work. The majority of construction firms understand that the better they are at their specialty, the more successful they will be. Our industry is highly focused on the technical and production side of the business. This is particularly true with small and mid-size contractors who often learned their trade before entering management or becoming the boss. Up until a few decades ago that was the only way to become a contractor. As the construction industry expanded rapidly after WWII, colleges and universities began to offer two- and four-year degrees in construction education. This was a giant step in the advancement of knowledge, sophistication, and level of respect for those employed in construction.

Construction people, who were sometimes referred to as *hard hats*, were on their way to becoming *construction professionals*. In addition to required courses in general education, construction schools offered technical and engineering-type courses. That left limited room in the curriculum for much accounting or business education. Limiting accounting and business courses seemed to fit well into an industry that already considered *production of the work* as their prime objective. The majority of today's construction industry leadership is comprised of people who came up through the ranks or who attended two- or four-year construction schools. All are skilled or educated with the technical information necessary to succeed at building the work.

Neither of the groups described above was exposed to much, if any, business education. Few had room in their curriculum for courses in accounting, finance, economics, management, marketing, human resources, business law, or the like. We are in an industry that seems to embrace the belief that if you can build the work, you will succeed. This may be partially responsible for our industry having little if any awareness of the necessity of accounting, finance, and business knowledge for the success of any type of business. Business is a science, and it is separate and apart from the product or service an enterprise offers. Business principles and processes are the same for mining, manufacturing, construction, etc. In the current competitive construction industry, it is critical to understand that there are two separate and unique aspects to every business: the production side and the business side. The business side is similar for all types of enterprises, while the production or service sides are different among types of enterprises.

There are several objectives of this chapter:

- Persuade readers that there are two separate and distinct sides to a construction enterprise: the business side and the production side.

DOI: 10.1201/9781003500315-9

- Convince readers of the critical role effective financial and business management has in the success and stability of a construction enterprise.
- Explain how to infuse these business skills into an ongoing construction enterprise or to expand and improve them.

There is some debate about the size of the role business management skills play in the success of a construction business. Many say *if you can build at a profit, that's all you need.* I hope by this point that I have opened the door to the reality that business skills are a component of a construction company's success and stability. The industry is populated by primarily type A personalities with technical or engineering talent who will likely ask, *how much business skill are you talking about*? Some may argue that more time and effort is spent on the production side than on the business side; therefore, production matters more. I suggest that the production side will not prosper without the business side, so you can't have one without the other. Production and business abilities are of equal significance to success.

Let's also consider this certainty: Serious mistakes on either the production side or on the business side of a construction enterprise can (and have) cause(d) financial loss, distress, and worse—insolvency. For that reason alone, it seems that the technical and business sides are of equal importance. This is not to say production is not vital. It is simply not more important to survival.

In his book *Buy Low Sell High*, Dr Dick Levin says that the title of the book is all you need to know about business. He goes on to say, no matter what a business produces or sells, its purpose is to make a profit. In his class at the University of North Carolina, he was asked: *If buying low and selling high is your definition of business, what do you call a company that is losing money?* Dr Levin answered: *I'm not sure what to call it, but you can't call it a business because a "business" makes a profit.* Dr Levin liked to clarify that businesses, other than non-profit companies, which he referred to as charities, are for-profit enterprises. He made clear in his teaching that what a business produced, bought, or sold was its occupation. How it functioned financially was its "*business*." Dr Levin clearly separates the business side of an enterprise from what it produced.

Measuring Stability

The business side of a construction enterprise has another distinction. It is the primary cause of construction business failures (Figure 8.1). The primary cause of financial distress does not originate on the production side. This generally surprises construction people. As I have said already in the text: None of the thousands of failed construction companies in my database failed because they did not know how to produce the work. All of them had been in business and producing the work for a minimum of ten years, and the majority had been in business for more than a generation. These firms no longer exist because of their inappropriate business decisions or inadequate accounting processes. Their record keeping failed to expose underperformance in time for anything to be done about it. What is frightening about these failed contractors is that all but one were surprised, even shocked, to discover they were in financial distress or insolvent.

In the previous chapter, it was demonstrated HOW accounting deficiencies masked losses from discovery.

In this chapter we will be learning WHY such accounting and reporting deficiencies and weaknesses were able to occur at all.

It takes hard work to be successful in the construction business, and the primary prerequisite is the ability to produce the work. It takes a long time to learn and master those skills, which leaves

Figure 8.1 Most construction firms excel at their technical specialty, yet business skills often lag behind.

little time to also take the many courses included in a business education. Many in the industry like to claim you can learn *how to manage a business* as you go along, which, if you give it any thought, is simply not so. You cannot assimilate complex subjects like economics, finance, accounting, statistics, and many others. They are just too complicated to be instinctive and simply cannot be spontaneously envisioned as you go along. Few, if any, contractors, particularly in small or mid-size firms, come into that role with any formal business or accounting training. The same is true for the majority of construction executives. They are construction experts, not accounting/business experts.

Some contractors have argued that their in-house accountants have the necessary business, accounting, and management skills and therefore can provide oversight. This brings up a sore point with me. I have spent the better part of 40 years trying to convince contractors that it is critical that their in-house accountants be active members of their senior management team. They aren't listening. Industry-wide there has been minimal acceptance, if any, of this position. Several contractors have tried to convince me that their accountants are part of their senior management team and that they participate and are listened to.

My test has always been to ask where their accountant's office is located. In every case the accountant's office was nowhere near the center of power. In some cases, the accounts office was not even on the same floor, and in other cases, not even in the same building. My next question is always: Does the accountant participate in all senior management meetings? The answers are usually something like: *Accountants don't know anything about construction*. To which my response is: *So what. Your construction people don't know anything about accounting*. Some contractors argue: *He/she is not really an accountant. He/she is a bookkeeper.* Then hire an accountant!

One of the reasons for the title of this chapter, *Industry Business Competence is Weak*, is because industry in-house accountants are underutilized, not trusted, and not listened to. This reality is a shortcoming in our industry because every other industry in the nation openly relies on their in-house accountants, and many CEOs of huge businesses were formerly accountants and/or earned a Master of Business Administration degree. Businesses that are too small to have an in-house accountant generally utilize their outside accounting firms. In construction we generally use in-house accountants for record keeping. The core issue seems to be that the majority of contractors believe (and I have been told this by many): *The only reason we have an accounting department is to make payroll, to comply with state and federal tax requirements, and to handle things like insurance and other paperwork requirements.* Our industry has accountants. We just don't know what to do with them. This explains in part the title of this chapter: *Industry Business Competence Is Weak.*

Reality Check

I genuinely hope that at this point most readers will agree, at least for now, that a construction enterprise has two distinct functional sides—production and business management. I further hope that I have made a reasonable case that the importance or significance of each functional side is equal. If not, I hope readers will agree that the business side is at least relevant and must be dealt with. What may be more difficult to agree on is that the quality (even accuracy) of business decisions can be improved with knowledge and education in economics, finance, accounting, statistics, and the like. Because weak or inappropriate business decisions are the primary cause of construction enterprise failure, the introduction of these skills is not optional but critical. I am not suggesting that contractors go back to school to gain knowledge in these subjects; I am saying it is critically important to bring this knowledge into management decisions and strategies of the organization. Best of all, this knowledge is already available in construction organizations that employ competent accountants. We just need to use it.

Unfortunately, too few of our CFOs are included in top management meetings and decisions. Most contractors that I talk to about this counter with an assurance that they meet regularly with their CFO. These are almost always one-on-one meetings, which inevitably are just *reporting sessions* with the contractor evaluating the information and telling the accountant what to do. When I point out that such meetings are clearly not to seek input or advice, the typical contractor response is: *Of course not—my accountant doesn't know anything about construction.* This just validates my position that in our industry, accountants and CFOs are not listened to or included in top management decisions. Some will point out that there are exceptions, but not enough to matter.

We are managing businesses in the industry with the second-highest failure rate in the nation. And we are doing it without any guidance from the sciences of economics, finance, accounting, and statistics.

One contractor countered: *After all, we're not a think tank!*

On this issue I feel like I am fighting an uphill battle because I am proposing a simple solution to a problem our industry refuses to even admit exists. We produce the entire built environment admirably because we accomplish the operations side of our businesses excellently. We do not accomplish the business side of our business well enough to maximize potential earnings; or to minimize financial risks; or to reduce the rate of business failures. And what makes it all worse is that no one can solve a problem that few believe even exists. Fortunately, there are solutions, and I am hoping that a discussion of the solutions will assist in clarifying the problem.

What to Do

The solution is not complicated. Recognizing the need is half the battle. Most construction organizations of any size already have a competent CFO who has education and training in economics, finance, accounting, etc. (Figure 8.2). If your CFO is not strong—there is no other way to say this—they need to be replaced. Smaller enterprises may be able to rely on their outside accounting firm. Contractors need to understand that your CFO does not need to know anything about construction, any more than your top construction people need to know anything about accounting. They are already qualified to participate in top management where they belong. If they are not qualified, that is your problem, and you need to solve it. Successful construction contractors who want to remain successful or improve their success will advance their CFO to a senior position equivalent to the most senior operations person. To defend against the unrelenting industry failure rate:

- The CFO will be included in their top management group.
- The CFO will be at *every* top management meeting.
- The CFO will be respected as an equal and listened to.
- The CFO will be copied on all senior-level policy matters.

The CFO must be introduced by the contractor to the top management group as an equal member with specific financial skills, which are to be considered in all future decisions. Economics will not dominate decisions but, depending on the group's assessments, may influence directions. This is not intended as a major change in decision-making processes, but as an important added influence

Figure 8.2 Including financial expertise in top management decisions strengthens long-term contractor performance.

on decision making. Those that have advanced their CFOs have experienced improvements in overall corporate performance, profitability, and stability. Infusing economics, finance, and accounting into the top management decision-making process makes the effort more than worth it.

Managing Personality Differences

Contractors and construction professionals should understand that part of the reason they are in those positions is because they are typically strong-willed and self-confident. Accountants are generally more analytical, detail-oriented, and cautious. These personality differences should be discussed and understood by all concerned. One personality type is not better or worse than another, but the difference can cause misunderstandings. For example, one personality type (managers) may respond rapidly to questions while another type (accountants) may hesitate and think longer before responding. For type A personalities like construction executives, this is often a problem if they are not prepared for it.

Someone needs to help managers understand and anticipate the differences in conversation patterns and the reality that accountant types often hesitate before answering a question. It does not mean they don't know the answer. It means they think before they talk—something Type A people do a lot less of. I point this out because I have observed uncomfortable interactions with disappointing results. Personality diversity is not a burden but an asset in human interactions and improves outcomes. Like-minded people are prone to *group think* and lack checks and balances. Diversity is a solution, not a problem that our industry needs to appreciate.

We can't change the reality that one of the largest industries in the nation is made up of over two million separate organizations—with the majority being small and mid-size firms. But we can and must recognize what that results in. Most construction companies were founded by one or two individuals. Most small and mid-size firms have a single owner, and those with partners often have a majority owner. The result is that the decision-making power lies, for the most part, with one person or with a very small group of related or like-minded people. Consider the difference in responsibility between the average contractor/owner and their top managers. Even in the case where a contractor seeks input or advice from their top managers, the decisions are usually made by the owner(s) of the firm. While others may have input into major decisions, the results of the decisions have the greatest impact on the stockholder(s).

Something Needs to Change

The full weight of major decisions is on the contractor. And any participants in decision making know that. Over time, for whatever reason—perhaps human nature—contractors seem to become more independent. In my experience, most contractors who have been successful for any length of time make all of the major decisions. Some maintain a facade of input from others while saying, *I founded this business, so no one knows more about it*. And others explain, *My company's doing fine so why change anything*. This might be convincing except for the fact that we are in the second-highest failure industry. Something needs to change. The following are just a few examples of the industry-wide disregard for accounting reports and for accountants that I have observed.

Example One

A huge third-generation West Coast construction firm doing over a billion dollars of work annually failed. It was a closely held family business with a board of directors comprised of all insiders. The firm's bankers and surety representatives believed it was well managed, and from all appearances it was. The firm had a large accounting department, and several of the senior accountants were CPAs. After working with the firm for some time on behalf of the surety, I asked the CFO a question. In a private meeting I asked how the firm could have been in this much financial difficulty without him knowing it. The response was I did know it. My next question was, why didn't you advise management about the losses? His response was: *the former CFO did advise top management a year ago and was fired. That is when I was promoted to CFO.* He reported that management had told him that the projects in question just appeared to be losers but would do fine. He added: *Management made it clear that they did not want to hear from me on this topic again.*

My fear is that readers may think I am making this up. Other than I realize this may be hard to believe, I have nothing more to say about this matter other than it was consistent with my experience that in-house accountants are not listened to. I am simply trying to make the point that this was a highly qualified and credentialed accounting department—and top management had no interest in their accounting reports or what they had to say. I believe that readers need to understand the critical nature of the obvious lack of interest (even disrespect for) accounting information. Unfortunately, this is not uncommon in our industry.

Example Two

I was asked by an executive vice president of a large Midwest closely held construction firm (almost two billion in annual sales) to meet for lunch. He requested that what we discuss be strictly confidential. At lunch, he explained that he had been selected to become the next president of the company. He asked me to help him interpret the company's prior-year financial statement. I asked him if he did not understand financial statements. He said he thought he did, but that this was the first time he had seen his firm's statement and that he can't believe what it indicates.

I was shocked to learn that none of the executives in the firm, including him, had ever seen a company statement. I asked, in that case, how did you know how the company was doing each year. His answer was he was only told in a private meeting how his division was doing but never saw any paperwork about it, nor was he ever given any financial information about the overall company. What this executive vice president was having difficulty believing was that this fourth-generation firm had been losing money for the last two years that the financial statement covered. He needed me to verify that he was reading this correctly and then went on to seek advice about how to deal with this.

I was flabbergasted because financial information is a *report card* necessary for anyone to manage their portion of the business. This unfortunately suggests that the business-side information of even huge construction enterprises appears to be considered of limited value. It is obviously not recognized as equal to, if not more important than, the production side. My experience was that some small and mid-size construction companies did not share financial information, but I was learning that even jumbo-size firms did not. This was part of the learning curve for me as I went on to discover that the majority of construction firms did not share their financial data with their managers. This matter is part of the reason I decided to write this book. These are critical concepts that need to be exposed, explained, and rectified if we are ever going to reduce the industry failure rate.

Example Three

During the financial analysis of a mid-size Southwest contractor for a bonding company, I discovered numerous bonds for projects on the West Coast but no job records at this office. The answer turned out to be that the west coast projects were actually performed by another unrelated firm.

The Southwest contractor had been *informally* lending the firm's bonding capacity to a small West Coast contractor for a fee. Because the projects were small compared with the Southwest projects, they went unnoticed by the surety. This had apparently started with just a few jobs but had grown over the years into quite a number of projects. All the West Coast projects had stopped, suddenly and owner complaints to the bonding company were what triggered the investigation.

During a rapid evaluation of the inactive west coast projects, it was easily determined that the small contractor had passed away, so the few employees and several subcontractors stopped working. The Midwest contractor knew nothing about West Coast projects, did not even have a record of their location, and refused to accept any responsibility for the contracts. The bonding company was responsible for the cost of completing these projects and paying for overdue material and subcontractor invoices. Under the terms of the indemnity agreement, the Southwest contractor was responsible for reimbursing the surety for the substantial loss. The Midwest contractor's defense to the bonding company was that the problem had nothing to do with his own firm which, of course, did not hold up.

The liability and the loss of future bonding put the otherwise successful contractor out of business. The cause of this business failure was obviously not a lack of construction skills because this company was profitable. This organization knew how to build the work for profit. The cause of the failure was that the contractor's *business competence* was weak. This builder had lent some of his bonding capacity to someone for a fee without even knowing that he remained contractually responsible for the projects involved. This contractor confirmed that he had never read the indemnity agreement with the bonding company that he had signed and claimed he does not even remember signing it. Until his attorney explained it to him, he actually did not understand that he was liable for the completion costs.

Business Competence

Readers should be proud to be a part of the industry that has produced the global built environment. The industry has the proven capacity to construct. The same industry, by virtue of having the second-highest business failure rate, does not demonstrate a collective capacity to manage the business side of a construction enterprise. Contractors put the work in place for a profit, and in many cases for generations—until they don't. Some will argue that only some contractors fail, so the majority demonstrate that they have business competence, but not going out of business doesn't mean you are competent at managing a business any more than not having any accidents means you perform your work in a safe manner. Additionally, the statistics don't seem to back that up in that failures tracked since the end of WWII are spread throughout the industry. Every type and size of construction firm in the US is represented, with almost half being second generation or older and some third, fourth, or fifth generation or older. Several years ago, the oldest construction firm in the world failed with losses in the billions.

Being so randomly spread, these statistics confirm that type and size of company are not primary causes of failure, which confirms, at least for me, that the primary causes of construction business failures are on the business side of the organization. That would mean that the solution requires greater attention to the business, management, and accounting side of the organization. A three-year research project on this proposition came up with an overwhelming list of business-related responsibilities a construction contractor of any size will have. These are business issues every business faces daily or periodically and must be directly addressed skillfully or addressed by seeking expertise outside of the organization. Half the battle is simply being aware enough to know these critical topics exist and need to be addressed. In reading the list, consider the potential results if a contractor does not even recognize that they are responsible for one or more of these essential business issues.

Contractor Responsibilities	
Accounting	use of financial information/planning, cash flow management
Banking and Borrowing	use and management of relationships
Business Opportunities	evaluation, strategies to enter or get out of
Change Orders	management, pricing
Computer	technology, appraisal, selection, acquisition, usages
Contracts	negotiations, administration, understanding of contract law
Crisis	management, avoidance
Customer Service	client maintenance, customer relations, networking
Designer/Inspector	managing appropriate relationships
Disputes/Claims	avoidance, negotiation, resolution
Equipment	evaluation, purchase, tool selection, maintenance, management
Estimating	strategies including adjusting pricing to circumstances
Labor	management, negotiation, understanding prevailing regulations
Leadership	including employee selection and management
Marketing	advertising, networking, screening potential customers
Marketplace	understanding, evaluation, market evolutions and cycles
Motivation	team building, earning loyalty, including compensation, benefits/perks
Organization Structuring	maintenance and updating
Overhead	management and control
Planning	business, short-term tactics, and long-term strategies
Productivity	management, operating policies, production standards, costs
Professional Development	of self and employees
Project Selection	screening, market and competition analysis
Risk	recognition and management, including insurance evaluation/selection
Safety	management, administration, OSHA, all prevailing safety regulations
Subcontractor/Vendor	screening, selection, management
Surety	use and management of relationships

A number of contractors have claimed that they have people to do some or most of these things on this long list. In effect, all of the actual effort to accomplish just about everything on this list is done by someone. However, that does not alter the fact that the contractor/owner/CEO must know what has to be done, must assign the work, and must be qualified to evaluate the substance of the effort. Some contractors have said that they don't know anything about one or more of the items on this list, which is exactly my point. Business competence in the construction industry is weak. If both this and the failure rate are true, we have uncovered the problem. All we need to do is develop the solution. The starting point for that is to understand the personal *attributes* that contribute to business success, and it starts at the top.

Attributes of successful CEOs	
Ability to Listen Attentively	Candor
Generosity of Spirit	Assertive versus Aggressive
Realistic Self-confidence	Sense of Humor
Curiosity	Clarity
Honesty	Empathy (You can feel this)

Few individuals have all these attributes, and those who have some of them will have them to varying degrees. These are the attributes that research by psychologists and management consultants found prevalent in successful leaders. I have encountered the argument that these won't work in construction, with some claiming that "toughness" is missing. These types of comments were mostly from mature contractors. I don't deny that toughness was widespread in our industry some decades ago, having personally experienced it in how I was treated as an apprentice. I hope that there is no need to convince my readers that today's workforce will not respond well to toughness. In fact, the image portraying construction as a tough industry may be partially responsible for fewer young people selecting the trades as a career choice. Let's stick with what the data tells us.

If you go through the list one item at a time and consider the impact each attribute has on the behavior of a leader, then ask yourself if you would like to work for that person; then by the time you get through the ten attributes, you should be able to understand why they are on the list in the first place. Few people have all of these attributes, but those who wish to be the best can make a conscious and continuous effort to improve in every area. Self-improvement is a lifetime activity. This is not a leadership book, but there are plenty of those out there, which I strongly recommend for readers with such ambition.

I am often asked how do you hire like-minded qualified leaders and managers. For openers I don't recommend restricting senior managers to people like-minded to yourself—no more than I recommend family members over outsiders. I have nothing against family members, but I am against the supposition that family members are automatically qualified. The ability to manage a construction business is simply not hereditary. The selection criteria must be the most qualified. For senior positions, I engage an industrial psychologist to do the screening and to make the recommendation of the best candidate, or on occasion, the best two candidates.

There are all types of employment experts available. This can be costly and time-consuming, but it helps avoid the exorbitant cost of hiring the wrong person and having to start over again a year or so later. Most employment experts take multiple steps to test and substantiate the qualifications of candidates being considered. Such as:

- Qualifications: They test applicants and verify credentials.
- Experience: They verify experience with all previous employers.
- Reputation: They determine the candidate's reputation with peers, prior employers, designers, project owners, and others to the extent possible.
- Financial Status and Stability: They review FICO scores, credit bureaus, and other publicly available information.
- Background: They run a background check to uncover any civil or criminal history.
- References: They personally double-check all candidate-provided references from former employers, clients, customers, or others.
- Personality: Most professionals utilize some form of personality test or other standard test that they deem appropriate.

One of the shortcomings of this process is that it takes considerable time to do it well. For critical senior positions, it is well worth the delay. Some are concerned that the candidate may be hired by another while waiting for this process to unfold. The reality in senior positions is that candidates are already employed and are exploring options for moving up or are dissatisfied. I have never hired an unemployed senior executive. During the extended interview of the recommended candidate, which usually includes dinner with a couple of my key executives, I attempt to get to know the individual as well as the candidate—to the extent possible. I am not big on prepared questions, but I always ask him or her what their long-term ambition in the company is if they are hired. I am not particularly recommending this question. It is just something I use to explore the candidate's outlook.

Improve to Relax

Years of assisting underperforming construction companies to *turn around* their failing businesses verified that the necessary improvements were exclusively on the business side of the organization. Not the production side. There are two reasons for this. One: The business side is where problems can be fixed. Two: Of the very few companies I encountered that could not perform their specialty profitably. I had no choice but to advise them to liquidate. I had to explain that the ability to perform their specialty at a profit was an absolute prerequisite for their being considered a construction company. If you think about it, you can't actually be a construction company *learning how* to construct or *trying* to construct at a profit. There is no such thing. Companies are *for-profit* organizations.

Fortunately, all the companies I worked with are still in business, and that is because improvements on the business side, while very difficult for top management to see, are relatively easy to discover and correct. All the contractors I worked with were skilled at building their projects and preferred their firms to be continually expanding but complained about uncertainty about their financial and business side. After their business/financial issues were rectified, they all said in one form or another that they finally felt *secure* in their businesses and no longer *threatened* or fearful of the unknowns. All agreed that by understanding the business and accounting aspects of their company, they never felt more *in control*. Some said that the business side of their company was *transparent* for the first time. If this doesn't cause you to take the next steps, I don't know what will.

Accuracy of Record Keeping

You will NEVER be certain where you stand financially if you don't trust your in-house accounting reports. It is critical that your accounting systems:

- Capture 100% of the firm's financial data precisely on time.
- Record 100% of the financial data immediately and accurately.
- Produce 100% of the firm's financial reports on schedule.

This may not seem difficult, and I suspect most readers believe that their firm's accounting processes already accomplish these exacting levels of precision. I, on the other hand, am not certain it can even be accomplished under any circumstances. I say this because of the tens of thousands of construction company financial statements my firm has analyzed; we have never found one that did not have some liabilities reported in a subsequent month or months after the invoices were received or the liabilities incurred. As explained earlier, any invoice received after a firm's books are closed for a given month must be recorded in the following month. Such occurrences are considered by most to be outside the firm's control, which of course does not change the reality that the financial information, including job costs, is incorrect in both months.

There are many inappropriate or undisciplined internal invoice handling and approval processes commonly in use, as presented in an earlier chapter. This unfortunate common mishandling of invoices on the production side of the business delays the liabilities from being recorded in the accounting systems on the business side of the business. We can't blame our accountants for the delay. Accountants will always caution against any delay in recording liabilities. It is top management's responsibility to produce the work at a profit, and that includes creating, managing, and supervising systems that measure performance. And if the CFO is not included in ALL top management deliberations, that is not going to happen because they have knowledge and skills that most operations people do not. I don't know how many other ways to convince construction practitioners that we cannot succeed long-term without the inclusion of finance, economics, and accounting expertise. They are a vital component of successful construction operations. Accountants are not *score keepers*. They are *players* and must be included in the *huddle*. They must be included in developing and executing the plays, not just measuring the results.

Summary

Construction is a team sport. Earlier in this chapter, *Contractor Responsibilities* were listed. I am certain many contractors would say they could not possibly be experts in all those separate disciplines. The actual construction of a project may be able to be accomplished without accounting, finance, and economics expertise. Running any type of business cannot function without accounting, finance, and economics skills. This goes double for a construction business because of the complexity of the product and lengthy duration of the projects. Knowing how to build the work is critical, but it is only one component of success. Critical components of operating a business successfully include accounting, finance, and economics. Common questions like, which is more important or what is the percentage of each, are meaningless. The answer is you need 100% of each. The good news is that most construction firms already employ the necessary talent. These firms just need to bring that talent into top management and listen to it. This will be a major step towards reducing, and perhaps eventually eliminating the failure rate plaguing our industry.

9 Understanding Financial-Operational Capacity Is Critical

Introduction

It is not surprising that a common cause of construction company distress, even failure, is exceeding a firm's financial or operational capacity. Unfortunately, most construction professionals are not even aware that there are actual limits to both. Capacity limits, financial and operational, for a construction organization are separate but interrelated. An understanding of each is critical to business success because if a construction professional doesn't know limits exist, they are likely to exceed them. Financial capacity refers to an organization's collective skills and knowledge of financial management, including access to financial resources, necessary for sound financial behaviors and decisions (Figure 9.1). Operational capacity refers to an organization's collective skills and knowledge about how to provide their service and produce their product for a profit.

Part 1: Financial Capacity

As discussed in a previous chapter, few contractors or construction professionals, except for construction accountants, have much formal education in economics, accounting, or finance. This makes the discussion of financial capacity with construction experts (builders and constructors) a little more difficult and a lot more critical. There seems to be an assumption within our industry that the financial side of the business is less important than th production side and can simply be left to the accountants. Hopefully by getting this far into this text, readers have realized that the *measurement* of financial performance is the responsibility of those working on the production side of the business. Measurement of performance requires input from in-house accountants but is not the responsibility of in-house accountants or outside independent auditors.

The construction industry failure rate will never be reduced until the financial and economic awareness of builders and constructors is increased. They do not have to do the accounting calculations, but they absolutely must be able to analyze and fully understand the accounting output. They must also understand the economic and financial implications of the accounting output. Some contractors say that their in-house accountants understand the economic and financial implications of accounting information. That may be true, but the accountants don't make the *strategic* decisions that drive the organization. Contractors, who do not analyze nor fully understand the accounting output, nor understand the economic or finance implications, make the *strategic* decisions that drive the company. This is not necessarily the case with very large construction firms. But research data, and the failure rate, confirm that it is definitely true for just about all small and mid-size construction firms. An exploration of the issues will help.

DOI: 10.1201/9781003500315-10

Figure 9.1 Financial capacity defines how well a contractor manages resources, capital, and financial decisions.

Cash Flow

Cash flow management is a major business activity sometimes minimized or ignored in our industry. For non-financial and non-accountant readers, cash flow is the net amount of cash and cash equivalents entering and leaving a business. Cash flow is crucial because it determines a business's ability to pay its bills, invest in growth, and cover day-to-day operations. Historically the construction process creates cash flow difficulties for generals and subcontractors because both are expected to put a month's worth of construction work in place before even *earning* the right to invoice for the work performed. We are not talking about getting paid for the work, just earning the right to bill for the work. Most construction contracts state that work put in place in a given month will be invoiced at the end of the month to be paid by the 20th, 25th, or 30th of the following month (some even longer). This means that the work put in place at the beginning of any given month will be paid for almost two months later. And that is if the payment is made on time by the owner of the project.

The contract also states that payments are contingent on the approval of the designer, which has been known to delay the payment process further. It is not uncommon for the designer to disapprove of a portion, or all, of the payment requisitioned, which of course further delays payment. Once approved, payment is reduced further by the contract requirement for the retention of as much as 10% of the approved payment amount. *Retainage* has historically been described as a reserve or guarantee that the approved work is in fact in place in strict accordance with the project plans and specifications. For readers unfamiliar with the construction industry, this guarantee is on top of the payment and performance bonds usually required by contract. These are issued by a licensed surety firm, which is normally an insurance company. Each of these bonds, payment bond and performance bond, is issued in favor of the project owner in the amount of 100% of the contract amount, totaling 200% of the contract value.

The performance bond guarantees the contract will be carried out in strict compliance with the plans and specifications. The payment bond guarantees that all labor, materials, and subcontractor costs will be paid in accordance with the contract. If not paid by the contractor, then the surety is obliged to pay these costs. The protection of these two bonds is in favor of the owner but provided by the contractor. To secure the bonds, most contractors must sign an agreement to indemnify the surety. This means if the surety incurs a loss, the contractor is liable to repay the surety for any costs incurred. These two separate *guarantees* reinforce the critical nature of a construction firm's *financial and organizational capacity*.

If a contractor takes a project that turns out to exceed the firm's financial or organizational capacity, the contract must still be completed. If the contractor is unable to perform for any reason, the bonding company must step in and complete the contract, including paying for all materials and subcontractors. Any loss incurred by the surety is then required to be reimbursed by the contractor under the indemnity agreement, which is backed up by the construction corporation's and the contractor's personal guarantee. Some in the industry refer to retainage and bonds as double protection or as b*elt and suspenders*. This can, and often does, cause corporate and personal bankruptcy. This demonstrates how very serious a total understanding of financial capacity is to contractors. The nature of this risk should make it equally serious to all construction professionals because everyone's decisions potentially impact the success or failure of a construction business.

Negative Cash Flow

The entire construction industry suffers from negative cash flow. A simple demonstration would be a one-person startup with a single one hundred- and ten thousand-dollar project (Figure 9.2). Costs are estimated at one hundred thousand with a profit of ten thousand, with a ten-month schedule to complete. Ten thousand dollars of construction will be put in place each month, so the building will be $11,000, and there will be 10% retainage. Major materials will cost $3,000 a month but do not have to be paid for until the following month. The $7,000 cost of payroll and expenses must be paid as incurred, including general and miscellaneous expenses.

The first month the contractor is out of pocket $7,000 and has material invoices for $3,000. When payment is made, less 10% retainage, it will be $9,900 ($11,000–$1,100 = $9,900). If payment is received by the 20th of the month, the contractor is out of pocket for the prior month plus another three weeks of the current month or 75% of $7,000, which is $5,000 plus. So, by the time the payment of $9,900 is received, the contractor is out of pocket $12,100 and owes $3,000 for material. After paying the $3,000 for material and reimbursing himself the remaining $6,000, he is still out of pocket $6,000 and is not able to even invoice again for another week. Now what happens if payment is made in 30, 40, or 50 days, which is not uncommon? By the time the first payment is received, the second month's costs are already incurred, and material costs are overdue. The net effect is that by the time any payment is made, another month of costs has been incurred, and 70% of those costs must be paid as incurred. The total retainage amount increases for the life of the project and is then often difficult to collect.

In actuality, contractors are out of pocket from the start of a project for a minimum of six weeks, and in most cases closer to eight weeks. In many cases longer because of disapproved

Estimated Project Costs, Revenue, and Profit

	Total	Month 1	Month 2	Month 3...
Revenue	$110,000	$11,000	$11,000	$11,000
Cost incurred				
Materials	$30,000	$3,000	$3,000	$3,000
Labor (payroll)	$70,000	$7,000	$7,000	$7,000
Cost incurred, Total	$100,000	$10,000	$10,000	$10,000
Profit/Loss	$10,000	$1,000	$1,000	$1,000

Project Cash Flow

	Day 30	Day 50	Day 60...
Amount Invoiced	$11,000	$0	$11,000
Cash paid out			
Materials	$0	$3,000	$0
Labor (payroll)	$7,000	$5,000	$7,000
Cash paid out, Total	$7,000	$8,000	$7,000
Cash received	$0	$9,900	$0
Cash Deficit	($7,000)	($5,100)	($12,100)

Figure 9.2 Example of a profitable project with negative cash flow.

invoice amounts or disputes. In this example we have yet to consider the overhead costs to operate the business, which include the office and the contractor's salary during the ten months. This contractor will only survive the cash flow struggle by going into debt to fund it. While this simplistic example demonstrates the industry's negative cash flow reality, consider the impact of multiple projects in various stages of completion. One method contractors employ to ease the burden of negative cash flow is a practice known as *front-loading*, which is where the contractor invoices for more than what is appropriate in the early parts of a job. This practice has become more difficult over time, and when it was more prevalent, it only assisted cash flow during the early stages of a project. The downside of front-loading is that it results in negative cash flow in the later stages of the project.

The reality is that all startup construction companies suffer from negative cash flow (Figure 9.3) and growing debt that will take years to reduce. Most startups fail after five years or more. The lucky ones make it to the critical 10-year mark. The failure rate drops dramatically for construction companies over 10 years in business. It is unfortunate that the startup failure rate is not widely known because people might resist the temptation to attempt a startup or approach it with greater caution. Part of the problem is the erroneous but generally accepted belief that contractors make a lot of money. Anyone taking the time to collect the statistics would discover that construction is a fairly low-profit-percentage industry.

Figure 9.3 Negative cash flow is built into construction contracts and must be managed to avoid crippling debt.

Capital Needs

Negative cash flow combined with low profit margins dramatically impacts financial capacity. To function in a negative cash flow business, liquidity is critical, which requires significant capitalization of the business, proficient financial management, and secure banking relationships. To manage all this, a prospective contractor must know how much it will take to fund their startup business at various sales volumes, which may fluctuate during early years and during growth spurts until stability can be achieved. It should be obvious that during startup, every new project will likely require additional capital. What may not be clear is that the initial capital will never be freed up because it will need to remain in the business. Startup contractors seldom anticipate this, which is part of the cause of the high failure rate. Numerous contractors discontinue startup attempts when they realize that they must continue putting money into their new business, when they thought they would be taking money out of it.

Contractors that make it through startup will eventually experience an increase in borrowing capacity, which may allow some of the startup capital to be replaced with debt. This can continue unless the growth rate exceeds 15% at which point cash flow will go negative again. Additional capital or debt will be required. If a startup survives the critical 10-year mark, it means that they did not exceed their sustainable growth rate. Those that failed exceeded either their financial capacity or their operational capacity or both. Failed startup contractors clearly did not know that there were capacity limits, or they would not have exceeded them.

Sustainable Rate of Growth: Successful construction firms growing at a rate of 15% a year or less should be experiencing positive cash flow and therefore be able to reduce debt. Depending on how long, they remain at or under that growth rate they may get to a point where they have no debt at all. The only question remaining at that stage is financial security, which usually means some *financial reserves*. The topic of financial reserves or capital needs is controversial, and some will argue that it depends on each individual's tolerance for risk. The fact that profitable projects vary considerably in their percentage of profit earned, along with the unfortunate reality that even consistently profitable firms have occasional losing projects, confirms the high-risk nature of our industry. Construction is also a cyclical industry with the risk of a market downturn. Since WWII, the construction market in the United States has cycled into negative growth nine times. Fortunately, these downturns lasted only one to three years. These realities, along with the industry's consistent high failure rate, should influence decisions about financial reserves/capital needs.

Financial Planning

Contractors don't seem to spend much time on financial planning or determining capital needs. The financially distressed firms in my database found themselves in the hands of their sureties because of negative cash flow that they never saw coming. The analysis of tens of thousands of construction-business financial statements exposed that there was no consistency in capital management. There were firms with two or three times more capital than was being used in any given year; firms that had bank debt of close to 100% of their annual capital needs; and everything in between. Prudence dictates that a firm holding more capital than is being used should not leave the excess capital in the company where it is at risk. Prudence also dictates that firms borrowing to their limit are at serious financial risk should they experience unforeseen losses or any borrowing restrictions.

The amount estimated to operate a construction business in the following year should dictate the amount of money left in the business that year. For most firms this may vary considerably from year to year, affected by earnings and the amount the owner(s) takes out of the business. Through discussions with their internal or outside independent accountants, contractors should be able to get assistance in determining how much capital will likely be needed to manage the business in any given year. To aid this decision, you may want to know how much capital was in continuous use the prior year. It is also helpful to know what the highest, lowest, and average amounts used were over the previous five years. This information can provide some guidance in determining what capital needs will likely be in the coming year. It is prudent to also include some reserve for an underperforming year.

Need for Financial Reserves

A construction firm can have a difficult year for any number of reasons. For example, if all of a firm's projects were profitable, but half made less than anticipated, the firm may have a loss for the year if overhead was not fully covered. We are in an industry where unanticipated declines in sales reduce profits or cause losses more commonly than most appreciated. The frequency of marginal and inconsistent profit rates strongly suggests that construction firms should maintain some measure of financial reserves. Some in our industry will disagree with financial reserves, insisting that all capital should be *working* at all times to the extent possible. They define non-working or unused capital as *waste*. Some insist that lines of credit can be used for unforeseen capital needs. The concern about that is it's easy to borrow money when you don't need it. Even with confirmed lines of credit, credit grantors sometimes want to know *why the need to borrow*. This can be awkward when the need to borrow is to cover operating losses.

The question is always how much financial reserve is appropriate or prudent. It is difficult to quantify an unforeseen deterioration in profits. Obviously if a deterioration in profits was seen (anticipated), it would have been prevented. The difficulty with *unforeseen* problems is not just that they are not easily identified and they often take some time to fix. A case in point: *A new estimator brought in a few losing jobs, but by the time it was discovered several other losers were under contract. This translated into two consecutive losing years.* Unfortunately, this firm is no longer in business, so there was no practical amount of financial reserves that could have saved it. This reinforces the high-risk nature of the industry and that cash reserves are not a luxury or even optional—they are critical.

Every construction company should be prepared to survive reductions in profit and modest losses. In an industry where low profits are the norm and overhead costs are ever increasing, the risk of something going wrong is heightened. Unfortunately, the longer an organization is successful, the more contractors begin to relax and forget that they are working in an extremely high-risk industry. They forget that second-, third-, and fifth-generation construction firms have failed, and many begin to blame *circumstances beyond their control* and not management. The risk is constant and not just the risk of loss but the risk of bankruptcy. We don't need to be afraid—just vigilant, cautious, and mindful that the number of decades or generations in business offers no assistance in unforeseen circumstances.

Financial reserves, while debatable, are at a minimum prudent and, in my opinion, worth the minimal cost because of the *health benefits*. Financial reserves reduce stress and worry, which are indisputably health benefits—not to mention psychological benefits. The amount of financial reserve is optional, and I do not press the issue other than to say I am convinced it should be discussed widely within the organization and considered a cost of doing business. My suggested starting point or minimum reserve is an amount of reserve equal to or a portion of the amount of anticipated overhead cost that must be paid irrespective of earnings. To be more specific, this means that the financial reserve amount is the same amount as the total overhead expense for the year. Put another way, reserves might be the same percentage of total sales as the overhead percentage of sales or whatever makes management comfortable. Reduction in the failure rate depends on it.

Construction is a growth-driven industry, and managing the rate of growth is an issue that does not sit well with contractors. Many are convinced that the only constraints to growth are the marketplace and bankers. The "*we can build anything*" attitude seems to be accompanied by a belief that "*we can build as much work as we can capture.*" There is an absolute limit to rate of growth in the construction industry, and it encompasses both the financial and operational capacity of an organization. The 15% growth limit was explained in an earlier chapter. It has been addressed here under financial capacity because the data used to determine the 15% limit was financial data. However, the rate of growth in construction has both a financial and operational component. The latter will be addressed in Part 2 of this chapter.

Part 2: Operational Capacity

A construction company's *capacity* is the organization's proven ability to perform specific services and produce specific products for a profit. Operational capacity has to do not only with management's ability and experience but also with the level of knowledge and experience of crew members, crew makeup, and crew leadership. Operational capacity refers to a construction organization's collective skills, knowledge, tools, and equipment used to produce their product and service for a profit. As mentioned earlier, financial and operational capacity limits are separate but closely related. Financial capacity issues are almost identical for general contractors and subcontractors, and no differentiation was made in Part 1. The same is true for operational capacity except for certain general contractors, almost all of whom are construction managers and design builders who do not self-perform any work but rather manage the overall projects. They act in a consulting role and are therefore not included in this context as *contractors*.

It is appropriate to clarify that all subcontractors and general contractors specialize in types of work, which are much broader than the general categories we tend to refer to them in, such as plumbing, roofing, etc. For example, a plumbing contractor may specialize in residential projects or commercial projects or new construction or repair and renovation projects. A roofing contractor may do only flat roofs or shingle roofs. Contractors are highly specialized. Some may offer more than one specialty, but seldom do they perform every possible category of "roofing" work in their designated field. This is significant because many contractors get into trouble when they deliberately or sometimes inadvertently take on work that is in their field but outside their specialty. A simple example would be a contractor specializing in hung tile ceilings would not likely be qualified to take on a plaster ceiling. I once had to manage a project where a fire protection contractor specializing in copper wire fire detection systems inadvertently bid on and was awarded a contract to install a pneumatic copper tubing fire detection system. It bankrupted the firm.

Estimating

The first step in understanding operational capacity is to recognize that construction organizations are far more specialized than their trade designation. As explained above, identifying a business as an *electrical contractor* does not indicate what the firm does, which could be wiring houses, underground electrical distribution, high-tension wiring, and so much more. The same is true for every trade. Construction contractors are so much more specialized than even some people in the industry realize. Some have actually said: *We can build anything*, which is never true. That statement should always be more complete, as in: *We will build anything that we can make a profit at*. I realize that I keep repeating this, but it bears repeating. In business, the question should never be: *Can I build this project* because it is incomplete and inappropriate. The valid question is: *Can I build this project for a profit?* And even then, a contractor can never be one hundred percent certain they can deliver a project profitably until the project has been completed.

It is appropriate to remember that we are in a high-risk business of pricing our work in advance and continually trying to prove that our estimates were correct. Securing the work is the initial activity on the *operations* side of the business, and we can only become proficient at estimating the cost of the work with practice. That means pricing the work based on what it cost us to produce previous similar work, which is the problem. A startup has not produced any work (for a profit), so it might be described as not qualified to price any work. They are forced to learn by trial and error, which is a major stumbling block even with previous experience working for someone else. A startup has no direct experience in selecting pricing and producing a project themselves, though many will claim that they did all that for someone else. Which, of course, is not the same as being at risk for each step, which is verified by more than half of all startups failing within their first 10 years. It is a "Catch 22" dilemma. Marketing and estimating in construction are for the most part learned by doing. For those who may have had specific education in this area, it is perfected by doing. Either way, there is serious exposure for startups and continuing modest risk for ongoing operations.

Marketing

Historically, construction in the United States was exclusively a price-competitive business, so the industry has a limited legacy in marketing. With an increasing amount of work moving towards multiple forms of quasi-negotiated contractor selection, there are the beginnings or some "marketing" efforts. These are primarily efforts to enhance a firm's image and are, for the most part, limited to large and, to a lesser extent, mid-size firms—mostly general contractors. Trade contractors remain primarily price-competitive, so to the extent they engage in any marketing, it is usually directed at name recognition and image building. Small and mid-size general and trade contractors seem to take a *passive* approach to project selection, preferring to wait to see what projects come into their marketplace and then work at getting onto the bidding list for those they are interested in. This process causes some firms to go after projects that are available but not necessarily within the organization's proven specialty. While these projects may seem like the only work available at the time, moving outside an organization's proven profitable specialty increases the risk of successfully performing such projects to the same level of risk as a startup.

Waiting until projects are put out to bid is too short of a horizon, but I have been told by contractors that waiting is the only way to find out what is available. The reality is that projects out to bid were in the design process for many months, even years. And prior to that, the owners of those projects were considering their needs, deciding on locations, and selecting designers. Projects are in the planning and design stages a year or more before the construction stage. For example, a large contractor client asked how they could do any long-term planning when they never knew what federal work they specialized in would be available until the jobs were put out for bid. I explained to them that the federal budget is a two-year process updated every year and that it is in a public document that includes all projects that are funded or awaiting funding. The client was skeptical, and I suggested they call one of their federal clients and ask about it, which they did. Their federal client verified the two-year budget process and offered to send a copy to the contractor.

This contractor now knows what type, budgeted cost, and location of projects will be coming out well in advance. This information eliminated the worry about available work and also prevented the contractor from going after less attractive projects, thinking that was all that would be available. This *public* information had been available to my client's grandfather and father, who ran the firm for the two prior generations, but they did not know it existed. Some contractors who work for state and municipal government agencies are unaware that these also have public budgets and project information available. In many cases, privately held and public corporations make no secret of their future construction or renovation plans. I have seldom encountered a client that could not discover well in advance what construction activities their customers were anticipating.

Production

Introduction: Contractors may believe they have financial limits imposed by banks and bonding companies, but the majority do not believe there is a limit on how much work they can produce. Most are convinced that "*we can build anything*," and that the owners and designers are the only constraints. Some owners and designers require prior experience on the *size* and type of project. Few contractors seem to realize that owner/designer constraints based on prior experience actually prevent numerous contractors from overextending their capacity. As these may be new concepts for many readers, an exploration of *operational capacity* is in order.

Most construction companies are probably more *specialized* than they recognize, and the measure of their operational capacity is unique to each. Operational capacity includes both management

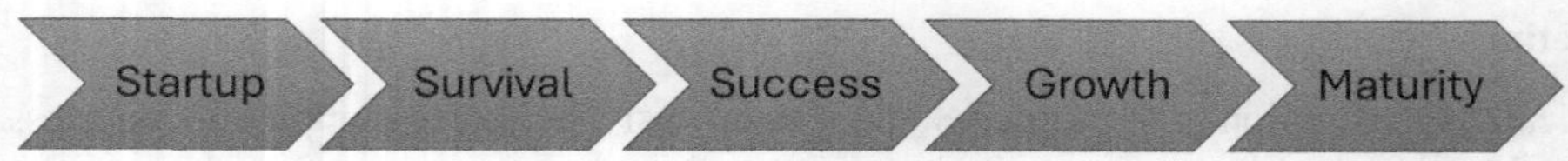

Figure 9.4 Stages of growth for a construction firm.

and crew-level experience and the quality and quantity of available tools and equipment. Management and crew experience with the type, size, and location of a potential project is a critical ingredient in potential profitability. Subtle differences in the type, size, or location of project experience can alter results considerably. We are not talking about startup firms here because they will encounter differences for many years until they stabilize and discover their specialty. The discussion of operational capacity here involves firms that have advanced from the startup to the stages of survival, success, growth, or maturity (Figure 9.4).

Type of Project

This discussion of operational capacity does not address obvious changes like experience with a single-story building vs a high-rise building or a road project vs a bridge project because these would challenge any firm. The industry needs to understand that subtle changes in the types of work attempted are a primary cause of contractors' financial distress and even failure. Most construction professionals don't even recognize the subtle changes as differences at all. For example, wood framing to metal stud framing, or a standard reinforced concrete slab to a post-tension concrete slab, or underground gravity sewerage piping to underground forced main sewage piping are changes contractors make regularly with disappointing results. The list is endless, but the point is that these slightly different types of work are performed by the same trades but require different skill sets. On most of the projects that failed because of changes in the type of work that my firm had to intervene on, the contractors said they did not see any differences in the work when they priced it. However, just as many said they did realize the difference but wanted to expand into that type of work. In either case they lost enough on the project to threaten the survival of their business.

It is critical that company owners and construction professionals of any size firm understand that they are more specialized than they realize. Our industry lists 16–18 different construction trades, so most do not see themselves or their organizations as specialized beyond their designated trade. A plumbing contractor, for example, who does only residential work does not call their firm a residential plumbing company. An electrical contractor who does only high-line electrical distribution wiring does not call themselves a high-line distribution contractor. Each, however, needs to be aware of their specialty, or it might lead them to do work they have no experience with, which is equivalent to a startup attempting their first project with no direct experience. And like a startup, the experienced plumbing or electrical contractor would have no direct experience to call upon to accurately estimate or proficiently construct work they had never done before (Figure 9.5). They may be able to complete the new type of work, but completing the work is not the objective. The objective is to complete the work for a profit.

I have been approached on this topic with comments such as *how else are we going to expand into a new type of work.* My response is that the statistic for entering new types of work shows that an excess of 50% will fail to earn a profit on new work. There is no known method for predicting how long it will take before earning a profit when performing a new type of work. Or perhaps more importantly, how much will it cost? Does that mean never do it? That is a contractor's choice, but the first guideline is: Understand the risk that more than 50% lose money on the initial try—and the amount varies from a little to a lot. What would a gambler or odds maker say—*bet only what you*

Figure 9.5 Operational capacity depends on experience with specific project types, not just a trade label.

can afford. If the potential loss or gain is unknown, your best guess will have to do. Starting with a large project seems unwise. A smaller project lowers the risk of the unknown exposure, which in itself is a huge step. If both the amount and the odds of loss or gain are unknown, prudence in projecting each would seem appropriate. The question is, can you afford the loss if the worst-case scenario occurs?

If the answer is no, obviously don't do it because you are, in effect, *betting the company*. If the answer is yes, there is one more issue to confront. History tells us that after an initial loss attempting a new type of work, it may take a few more attempts to turn it around to become profitable. The losses will usually reduce in subsequent attempts, then the second question becomes if you can afford the unknown but estimated initial loss, can you afford two, three, or more lesser losses added to that? If the answer is no, do you still want to make the initial attempt? At this point some argue, *what about the upside possibility that the initial attempt is profitable?* That is a valid question, but it is a gambler's question, not a businessperson's question. It is what happens when a business owner forgets that business is not a game. And it certainly is not a game of chance. Management avoids risk to the extent possible, and it is occasionally valid to accept a risk attempting to make more money even if it includes a small risk of making less money. However:

It is never advisable to engage in any risk of losing money.

This should not be up for debate because a business is, by definition, for profit. Businesses take risks when they do something to make money, but the *risk* in question is how much money that something will make. Businesses do not do anything where the risk is to make money but includes the risk of losing money. That is not *business*. That is clearly defined as *gambling*.

Learning from History

While it is possible to be forced back into the survival stage, no business owner or manager would ever deliberately expose their company to anything that even approaches the risk of loss. Businesspersons would therefore never deliberately, under any circumstances, expose their company to a loss of any amount. This should not be a hard lesson for those of us working in the industry with the second-highest failure rate in the nation. However, every construction professional seems to be convinced that *it can't happen to me*. The majority insist that the contractors who failed were simply not qualified, which is simply not the case. That is why only construction firms older than ten years or more are included in the failure database. All the companies included had proven success with more than half the data from second-generation firms or older—with many third or fourth generation and older. All the data referenced is from successful contractors of every possible size. If contractor failures occurred in multigenerational, huge, medium size, and small construction enterprises, they can happen to anyone. The only way it won't happen to any of us is if we learn from those who went before us.

Size of Project

Human nature tells us that *bigger is better*, so it is no surprise that contractors and construction professionals express a preference for continual growth. This is in spite of the published data that, on average, small projects earn a greater percentage of profit than mid-size projects, which in turn earn a greater percentage than large projects. It should be no surprise that jumbo projects earn on average the lowest percentage of all-sized projects. There is nothing wrong with growth, and it may be a *natural* entrepreneurial instinct, but few contractors seem to know that there are very real limitations to the rate of growth. Research has discovered that construction companies have a *sustainable* annual growth rate of 15% per year. That means that an increase of up to 15% in the dollar amount of work put in place (adjusted for inflation) can be accomplished while maintaining average previous profit margins. Construction firms that grew more than 15% did so at a reduction in profit margins. Average profit reduction continued to reduce through growth at 25%, at which point project losses began to appear. From that point, project losses increased as the rate of growth increased.

Successful construction firms generally have what I refer to as a *service mix*, meaning that they produce some small projects, mostly mid-size, and fewer large projects. Small, mid-size, and large are relative terms, meaning a large project for one firm may be a small or mid-size project for another firm. Some contractors refer to small projects as *nuisance work,* but small projects almost always make a greater percentage of profit than mid-size or large projects. Mid-size projects usually make up the majority of the firm's workload and can be described as what the firm does best. Some refer to these as their *bread-and-butter* work. These mid-size projects make a lesser percentage of profit than the firm's small jobs but are generally consistently profitable and hardly ever lose money. Most successful companies have a few large projects that almost always make a lesser percentage of profit than their mid-size jobs. However, these larger projects are needed to attain sales goals and to satisfy a desire to grow the company. These large projects are also how contractors usually identify themselves. These are usually the types of projects the contractor usually wants to be known for by prospective clients and the general public. Look at any contractor's website, and you will likely see pictures from the largest and most complicated projects that company has performed.

There are, of course, options on *how* to grow. These different growth processes are optional, but at the same time it is not commonly recognized that there are different growth processes. This likely means that the process of a company's growth is not selected by contractors but is probably a result of circumstances or opportunity. Any *change* in how a business operates involves risk, so it should come as no surprise that the common but not widely known growth processes each differ in their level of risk. The different processes of construction company growth are:

- *An increase in the number of projects-small, mid-size, and large*

If a contractor *increases the number* of small, mid-size, and large projects without appreciably altering the average size of the projects in each category, nothing on the production side has changed. They would therefore only be restricted by the 15% threshold on annual sales growth in dollars before risking a decrease in percentage of profit. At that rate there would be no change in the risk to the company. The reduced-risk 15% annual rate of growth is certainly not unreasonable, in that compounded at 5% inflation, it would double the size of the firm in five years, making it four times larger in ten years.

- *An increase in the size of small, mid-size, and large projects*

If a contractor increases the *size* but not the number of all of the small, mid-size, and large projects, there is a change in the production operations in that everything being built is larger. Therefore, the change in risk would be small but increasing because the change in average size of projects in each category means that some projects will be larger than the crews have prior experience with. If the increase in average project size in each category is limited to roughly 15%, the risk can be assumed to be minimum. The greater the increase in project size over 15%, the more the risk will increase. While there is no specific research on this category, it can be assumed that profits will continue to reduce through increases in the size of the project to 25%, after which some losses will likely begin to occur.

- *A decrease in the number of small projects and an increase in the number and size of mid-size and large projects*

The decrease or elimination of small projects abandons the highest profit percentage work, but the immediate financial impact is modest because small projects, while high in profit, are a minor portion of the overall profit amount. The abandoned small work is replaced by an increase of the firm's amount of revenue and profit from the mid-size and large category projects. Increasing the size of mid-size projects increases risks because this is the largest category of projects, described as consistently profitable, and as such is the work that the firm does best. Changing anything about what the firm already does best is a risk. Increasing the size of large projects increases risk not just because they are the lowest profit percentage category but also because, by definition, they are *not* what the firm does best.

Large projects were already the highest-risk type of project, so increasing their size can only amplify that risk. It may mitigate the risk by limiting the increase in the size of the mid-size and large projects to 15%. However, without the high-profit, small projects, there may be an overall reduction in earnings. To deliberately increase the size of mid-size and large projects by greater than 15% to make up profits missing from small projects will consciously and deliberately increase risk to compensate for management's decision to grow. This becomes complicated but hopefully highlights the purpose of this book: to bring to light the reality that constructing the work is only part of the business. Management, accounting, finance, economics, and statistics are equally critical parts in growth choices.

- *Elimination of small projects, a decrease in number of mid-size projects, and an increase in large projects*

Increasing large projects while decreasing mid-size projects and eliminating small projects is the highest risk of the four growth processes explored here. It also happens to be a popular growth process—particularly by contractors who become enamored with large projects. Reducing the higher profit mid-size projects with their limited exposure to loss may be considered imprudent. Particularly when this category of projects is described as: *what the firm does best*. Increasing the proportion of large projects, which are the lowest profit category of work and are already the highest risk category, obviously increases risk. An increased exposure is that it does not take too many mishaps for large, low-profit-percentage projects to slip close to break-even or worse. In such a scenario, that could result in a substantial impact on the overall company profit percentage. If two large projects were to go bad at the same time, it could be disastrous. With this growth process, both profit potential and profit security are reduced, which seriously increases risk.

Profit and Risk

There are some who will say that it is preferable to have a smaller number of large projects than a larger number of mid-size jobs, claiming that greater control over fewer sites is preferable. Others have argued that a focus on percentage of profit is not appropriate and that growth and size are all that matter. They also argue that a smaller percentage of a bigger amount is more profit. The point missed here is that any measurement of profit must include both the amount of profit and the associated risk. For example, a smaller amount of profit at low risk is superior to a larger amount of profit at high risk. Measurement requires both the amount of profit and the odds of winning. Or the odds of something happening or not happening. If someone gave you odds of five to one in your favor, would you take the bet? You can't evaluate that until you know the penalty or reward of the event. Suppose it was a game of Russian Roulette—a revolver with one bullet in it and five empty chambers pointed at your head. The penalty is too high for any rational person to accept these favorable odds, or any odds for that matter. There are some business decisions that never would have been made under any circumstances if the contractor had analyzed their decision in this manner.

Location of Project

Why does the location of a project matter so much? Because construction work is produced differently in different locations. How much different? Just a little, but in this industry, profits are just a little. Therefore, a contractor had better have an intimate knowledge of local conditions and practices before estimating the cost of the work—or attempting to perform the work in a different area. Clarification of this issue takes just a few examples of successful construction firms that were put out of business because they were not familiar with the local conditions in areas they attempted to work in. It should be no surprise that in most of these examples the firms priced the work in a manner they anticipated it would be performed. However, I discovered later that for one reason or another the work could not be accomplished in that manner in the location of the work.

In one case, a large successful non-union contractor contracted for a project in a location that they did not know was predominately union. The very large project was eventually finished by the bonding company. In another case a utility contractor located nowhere near the coast took an underground sewage project in an oceanfront town without realizing that there would be saltwater in the trenches dug for the work that would rise and fall with the tides. There was no pumping or water condition costs in the project estimate. Another contractor from a cold weather state took a two-year contract to build a huge concrete structure in a dry desert. There was no money in the estimate for the ice required to be used in the concrete mixed at the onsite batch plant that operated for eighteen months. These examples are all very large projects with disastrous results. Consider the small changes in how work is produced just a little differently within different areas of a region, state, or county. There are different customs, rules, and even products from one area of the country to another and even inside and outside of some city limits.

These different practices are neither good nor bad—some are simply customs and for the most part harmless. They are harmless to those who know them, those who are accustomed to the area, and those who know what they don't know. These differences are dangerous only to those who don't know that they may or may not exist anywhere and everywhere. I am confident that once readers understand this, they will find a way to tap into local knowledge or, better yet, think twice about expanding into unknown territory. Maximizing profit, not maximizing sales, needs to be our motto.

If growth and expansion in your own backyard increases risk, and you are guaranteed that it does, how can anyone possibly measure the risk of a local firm expanding into multiple states or nationally? Some may say we can, but my experience is that we can't. Years of research devoted to growth and rate of growth confirm that any growth greater than 15% a year has a risk that increases exponentially as the rate increases. The exposure of rapid growth for an already successful and profitable contractor of any size is simply not worth it under any circumstances. The only safe expansion is at the perimeter of a contractor's normal area or what I call the *proven area of influence.* The standard argument is, *what about all these huge regional and national firms?* My answer is the ridiculously high failure rate of our industry suggests that contractors of any size need to accept the reality that our industry has growth limitations. Simply look at the number of construction firms that might still be with us, had this data been discovered earlier.

It sounds counterproductive to say that the way to learn how to open multiple regional offices is open to one. But opening the first one successfully has such bad odds no one can recommend it. No one knew that rapid growth can and did cause the industry's failure rate. Now we know, but we don't want to accept it. A 15% growth rate is not a punitive limitation. No one likes limitations, particularly the successful Type-A personalities who believe they have already proven that the limitations do not apply to them.

Summary

The construction industry often equates success with the ability to build quality projects, but true success means building quality projects for a profit. Contractors must become as skilled in the business side of construction as they are in building the work itself. This includes understanding the very real financial and operational capacity limits that every construction organization possesses. These limits are not always obvious, especially to those who have built their careers on successful project execution rather than successful business management. Ignoring the critical need for business skills does not mean they do not exist. On the contrary, failing to recognize and respect these business constraints has quietly caused the downfall of countless contractors who never saw the warning signs coming. Knowing how to build is only part of the equation; knowing how much you can build, and at what cost to your company's stability, is what separates firms that endure from those that disappear.

10 Limits to Growth

Introduction

There are broader risks associated with growth than just exceeding an organization's capacity limit as presented in the previous chapter. Growth in the size of a construction enterprise and the *rate of growth* both increase the risk to an organization. Growth risk is exacerbated by the reality that contractors and construction professionals all appear to be fixated on top-line total sales growth rather than growth in bottom-line profits. Much like capacity limits, no one seems to believe that there are any *natural* industry-wide limits to growth in sales. Most construction people see any growth constraints imposed by sureties and banks as unfair, manmade restrictions on their rights to grow their businesses at any rate of growth they choose. The question that no one is asking is: *Are there actual growth risks in the construction industry* and *are there risks in growing an organization too fast?* The answer to both is *YES*.

The good news is that in a growing market there is plenty of work. However, attempting to do too much work can result in bad news. Growth is a two-edged issue, particularly in a cyclical industry. The overall construction industry has grown almost continuously since WWII; however, about every ten or twelve years it has slowed to negative growth for one to three years. Usually, with little warning, always with an unpredictable duration, and never without causing widespread reductions in profits accompanied by losses before bouncing back. At each market downturn, competition increases immediately as contractors have fewer projects to compete for. Prices come down, reducing profits and eventually creating losses. Price reductions and losses are directly proportional to the duration of the downturn.

Unfortunately, during the early stages of a market recovery, there is little or no improvement in margins. This is because everyone suffering the reduction in sales and profits desperately competes for the increased amount of available work. Profits eventually improve, but slowly and seldom rebound until the available amount of work approaches the size it was before the downturn. Complicating any recovery is the fact the labor market will have also decreased and available labor rebounds slowly. The problem shifts from not enough work to not enough labor available to perform the work. Historically, the industry has never experienced a time when anyone agreed that there are just the right amount of work and workers. Any market expansion should become a *seller's market*, but that does not happen in construction for quite some time after a downturn. The length of time to full margin or profit recovery is proportional to the length of the downturn.

Even though the length of the up-market cycles greatly exceeds the length of the down-market cycles, it tends to foster pessimistic and short-sighted market projects. During good years, contractor opinion polls often predict a slowing construction market, which is the same as polls taken during declining markets. One explanation may be that because the industry is fragmented with well over a million construction businesses in the country, contractors cannot easily get an accurate

DOI: 10.1201/9781003500315-11

overview of how the industry as a whole is performing. Industry consensus opinions appear to be mostly guesswork because they seldom match little known and rarely used market indicators. The lack of dependable market projections creates problems because the belief that there will be less work available causes aggressive pricing. The industry suffers from a lack-of-information problem rather than a market problem.

Margin Realities

While reduced profits may be expected in a down market, an unfortunate number of construction companies experience adverse effects from good markets as well. In a growing market some firms take on more work than they can efficiently handle and finance, suffering financial distress or worse. It is difficult to project the limits of how much an organization can effectively perform and finance. The surest way to determine limits is to exceed them, and even then, results may not be obvious for a long time. The incidence of financial distress in the industry is almost the same during market growth as it is during market reduction.

Rapid increases in the amount of highly profitable work put a strain on cash flow because growth increases cash outflow before any increase in cash inflow can occur (Figure 10.1). Include the cash withheld by retainage, and it becomes obvious that growth eats cash. Few construction organizations can gear up quickly enough to maintain profit margins during rapid growth periods. Hiring new people to accommodate growth impacts performance and profit. Crews having worked together on similar projects are often a key to profitability, and that advantage is easily diluted by

Figure 10.1 Rapid growth strains cash flow, people, and systems—often eroding profit instead of increasing it.

the addition of new hires. A rapidly expanding construction organization sustains increased risk even if closely and intensely managed. Both financial and production risks in construction are magnified during growth.

Overhead

Overhead costs are difficult enough for contractors to control when their companies are not growing, but in a growing organization, managing overhead is a very real and hazardous exposure. Because they can't add a half-person or a half-piece of equipment, companies are often forced to incur overhead costs during growth in larger amounts than they would like. This often creates losses until the size of the company grows into the overhead. The problem is magnified when it becomes necessary to increase sales just to cover the increased overhead costs, effectively putting the company into *double jeopardy*. Such maneuvers increase cash flow but do not necessarily increase profits. Particularly when it becomes necessary to reduce margins in order to capture the new work necessitated by the increases in overhead costs.

Pricing Dynamics

When an organization decides to grow at a rate which is greater than the growth rate of the overall market they operate in, it means the organization is increasing their market share. An increase in market share can only be accomplished by taking market share away from competitors. What few people stop to consider is that increasing market share beyond the percentage that their market is growing means that their competitors must reduce their market share. Many argue that is not their intention, but intention has nothing to do with this. The reality is if your market grew at 3% and you were able to grow at 10%, your competitors will have had to collectively reduce their sales by 7%. If they did not, then where did that growth come from? Growth by taking work away from competitors can only be accomplished with competitive price concessions.

Market share may not be a term or concept discussed regularly among construction professionals, and few make a conscious decision to lower their prices to increase sales. However, this is exactly what occurs. And unfortunately, when selling prices are reduced to capture more work, they are reduced to capture all the work, not just the *increased* portion of the work. You can't operate with two price standards at the same time. The inevitable result is that the company usually ends up needing even more volume than originally planned. This leads to a downward profit spiral because when an organization is stretched by growth, there is little time for anyone to see problems coming. More growth requires more overhead, creating the immediate need for even more growth. This chain of events has caused numerous construction business failures.

Measuring Performance

Rapid growth also puts a strain on a company's key people and systems, and sustained growth does not allow for a reasonable training period. Continual growth doesn't give an organization a chance to test new people or systems before the next wave of new people and systems is added. If performance deteriorates as a result of growth, it will only be discovered after the fact. Any corrective measures needed are more difficult to recognize or implement while managers and systems are stretched. During vigorous growth, overworked employees are coping with the largest volume the enterprise has ever handled, and supervision is stretched. Some companies can't recover from this scenario. Too many organizations pursue growth without measuring performance until it is too late.

Rate of Growth

Years of research indicate that growth for a construction enterprise of 15% or less annually is sustainable. A growth rate of more than 15% annually diminishes average profits, and any growth over more than a couple of years compounds quickly. During periods of continuous growth, the strain on people and systems is substantial—sometimes to the breaking point. Intermittent growth reduces risk. Just like a heavily loaded truck advances more efficiently up a series of modest grades than up a single steep incline, company resources perform better during short growth periods of a few years as opposed to being subjected to continuous growth over long periods. Continuous growth makes performance measurement difficult and measurement results unreliable.

Growth for a construction enterprise of more than 15% annually should be considered substantial, and it always increases business risk. Sustained growth over more than a couple of years compounds quickly. At a 15% rate of growth, a company doubles in size in five years and triples in seven years. At 25%, it doubles in three years and triples in five years, and at 50%, a company doubles in 20 months and will grow by 500% in just four years (Figure 10.2). Growth requires more resources of people, systems, and money. Performance depends on an organization's ability to find the necessary qualified people; put appropriate systems in place in advance of expansion; and finance the increased costs. Rate of growth impacts the likelihood that an organization will be able to bring qualified and adequate resources to bear on the new work in a timely fashion. After failures associated with growth, it was all too common to hear *if we could have found the right people, or had the right systems, or had the right financing, it would have worked.*

The alternative is to expect existing resources to do more; however, few construction organizations are known for having underutilized resources or *bench strength.* As volume increases, most people don't seem to realize that the *expanded company* is not the *same* company that it was and is therefore *untested* as an organizational unit. Many argue that it is the same company, just a little bigger. But as organizational functioning operates, when you add more than 15% new people, you change the *functioning* of the entire organization. The only way you can measure the new organization's performance and profitability is to wait a minimum of a year and test results. Testing any sooner is guesswork.

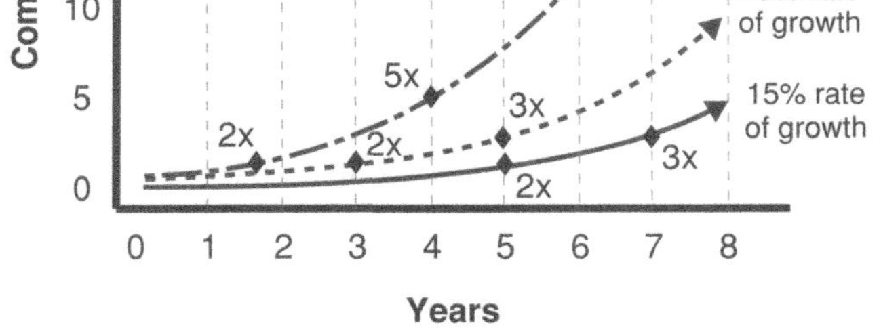

Figure 10.2 Company size compounds rapidly as growth rates increase, doubling and tripling much faster at 25% and 50% growth than at 15%.

With continuous growth, it does not matter what the test shows because with new growth being added during the year, the volume is never stable long enough to test. Unfortunately, it is impossible for an underperforming organization to roll back to their proven size and proven team, so testing is not a viable approach. That being the case, it may be difficult for readers to accept, but any growth over 15% annually is deliberately, if unknowingly, entered into without any advanced scientific measurement of its potential for success. I repeat:

Growth over 15% annually is entered into without any advanced scientific measurement of its potential for success.

Obviously, contractors grow faster than 15% annually all the time, confidently believing it will be profitable and/or to their advantage. The reality is that *believing, or being convinced,* or *feeling certain* is not the same as *scientific evidence.* Readers were warned that this text is a *no-holds-barred* report on the state of the construction industry and an attempt to determine the reasons for the ridiculous failure rate.

Continuous growth at any rate greater than 15% annually results in a new company that is unable to revert back to the pre-growth company because that company no longer exists. Many have *recovered* from rapid growth attempts, retreating to their former size, but they are not the *same* company. They may have also succeeded at continuous growth of more than 15% annually but they were at risk of failure during their growth, whether they recognized it or not. Determining the limits of expansion is complicated, but determining the risk of growth has been accomplished. There are respected management specialists who don't believe there is a limit.

However, there are enough companies that no longer exist, that were household names in the industry during their meteoric growth stage. This suggests there are indeed restrictive factors, or limits, to a sustainable rate of growth. While critics may point out other reasons for failures during growth, the reality is that rapid growth in-and-of-itself is dangerous—not always fatal, but always high risk. How is "high" defined? If the penalty is bankruptcy, then the risk is high. Just like Russian Roulet—because someone can pull the trigger and live does not mean it was not high risk.

Industry Cycles

Construction is a *lag* industry, which means it follows the US economy up and down but *lags* behind the US economic cycle by 12–18 months. Most people understand that the industry suffers in down cycles for obvious reasons that have been addressed above. Few people understand why many construction enterprises continue to suffer after the US economy has recovered and reentered the growth cycle. The construction of most projects has very long durations from conception through design and funding to the start of construction. This can take considerable time—on average 12–18 months—even longer.

When a US economic downturn occurs, economic activity slows, which for construction means that many projects in conception, design, or funding stages may be postponed. However, projects in the construction stage, having already been funded, usually continue to be completed. Therefore, production in the construction industry continues for another 12–18 months after the beginning of a US economic slowdown. The industry continues to prosper while contractors burn off their backlog, completing the work that was in progress at the start of the US economic slowdown.

When the US economy turns up, delayed projects that were in the conception or design stage begin to be released but will not be ready for funding and construction for an average 12–18 months. During this time following the improving US economic market, the construction industry continues to shrink in size. However, when the construction market finally turns up and begins to recover, margins or profits do not recover for another 12–24 months. This is the result of aggressive competition for the limited work available among contractors desperate to return to their pre-downturn size. The aggressive competition that limits margins continues until well after the construction put in place returns to the pre-downturn amount—or close to it. During market recovery, construction businesses are attempting to grow back to the size they were before the downturn. These are just more of the serious risks associated with growth and rate of growth (Figure 10.3).

If a downturn has lasted more than a year, firms will have been forced to lay off crew members and possibly managers. If during market recovery these positions are replaced with some or all new people, a new risk is introduced. Some work is being performed by untested tradespeople, and experienced crews are diluted with new people. Assimilating and training new people negatively impacts performance and therefore profitability. One might think that as the market rebounds, margins will rebound similarly, but that is not the case. Competition remains aggressive as contractors attempt to accelerate their growth back to their pre-downturn size. This is particularly true after an extended market downturn, as contractors continue to price work aggressively in an attempt to make up for the growth they feel they missed out on during the long downturn.

In these cases, the industry is its own worst enemy. Contractors who are generally considered optimists are often pessimistic when it comes to forecasting their market. Contractor *consensus* market forecasts are usually lower than the future market that develops. Admittedly these forecasts are more guesswork than research; however, believing there may be less work in the future can unconsciously impact pricing, which affects the probability of increasing margins. For many

Figure 10.3 Construction cycles lag behind the broader economy, creating hidden risks during both downturns and recoveries.

reasons, after several longer downturns, margins did not recover until well after total industry sales recovered to pre-downturn levels.

A word about consensus forecasting. The construction industry is fragmented, perhaps more than it should be, because each of the 18 or 20 major trades and a greater number of smaller, highly specialized trades have their own trade associations. And many trades have multiple associations separated by union, non-union, or in some cases size, specialty, or location. Most of these national and local trade associations poll their members for opinions on where they see the growth or size of their specialty work the following year. Thus, the term *consensus* forecast. Few forecasts look out further than one year. Each association publishes its own data, which adds some depth to the results. The result is the averaged-out collective opinions of a reasonable number of individual contractors.

I can address these surveys because for many years while in the contracting business with my brother, I participated in them. The opinions collected can't help but reflect the current circumstances of each contractor. For example, if we had just been successful at being awarded several projects, we responded with a good year ahead. However, if we were struggling to obtain work, we saw a poor market ahead. We relied on the consensus of our peers, but I realize now that statistically the results were just *opinions*. Opinions that, after being collected officially and reported in a magazine, appeared as fact. I am not sure how many firms use consensus data in their planning, but I would suggest that smaller firms would do better collecting data from their better customers and, to the extent possible, determine what their larger customers' plans are to forecast the future market outlook. As discussed earlier, the lower risk option of planning conservatively to the lower range of probabilities is more profitable in the long term than planning for the higher range when neither develops. Prudent risk management always avoids the higher odds.

Double Jeopardy

Research verifies that construction firms that are not growing or are growing at the modest rate of 15% or less earn consistent profits that will reduce over time with growth. We also know that statistically, on average, smaller projects earn a larger profit percentage than larger projects. Disregarding either of these realities reduces profit potential. The degree to which the rate of growth exceeds 15% or the project size is increased creates the potential for rate of profit reduction. These realities can be disregarded one at a time, but if a firm decides to disregard both realities at the same time, the reduction in profit potential is not just doubled but is potentially increased exponentially.

For example, if a firm grows at a rate greater than 15% annually but does not increase the average size and *service mix* of its small, mid-size, and large projects, they have disregarded only one of the two above realities. They will likely experience a reduction in profit percentages, which will vary depending on their rate of growth. (Service mix here means the ratio of small, mid-size, and large projects.) If a firm were also to increase the size of their projects at the same time, they would experience another negative impact on profits. The magnitude of the combined profit reduction is unknown, but experience tells us it would be exponential. The reason for this is that both changes require increased or new skills for the existing employees and for the new untested employees.

In spite of this, continuing industry failures indicate that a considerable number of small and mid-size construction firms continue to take on more work than they can efficiently handle and finance. None of these firms anticipate a problem, but unfortunately many of them encounter financial distress, and some ultimately fail. While information like this is not accessible to the general public, readers can be assured that this is actually happening as this is being written. There are currently a number of mid-size and large construction companies struggling with financial difficulties.

These are spread across the country, with a couple of older and more sophisticated enterprises included.

Because history repeats itself, we can be certain that a similar number of firms right now are considering decisions that will place them in this same category some years from now. It is interesting that an article in the news a couple of years ago about the failure of a huge construction firm included the comment: "*If they can get into trouble anyone can.*" These failures can be prevented, but not by construction practitioners unwilling to accept the fact that there is actually a need for change.

Growth vs Margins

The steady growth in the overall size of the industry over the last 15 years with little growth in the number of construction businesses means that those businesses are growing rapidly. This *seller's market* should almost automatically result in an increase in profits. Unfortunately, profit margins are not keeping pace. The industry is struggling with *growing pains* that always occur during continual growth without the benefit of increased profits, which usually occur during growing markets. What is happening is that overworked or new estimators and crew members are being led by overworked or new managers. This is a recipe for inefficiency—or worse disaster. Inevitably, some firms will grow beyond their organizational capabilities and become *casualties* of the robust construction market. The reason is because it is extremely difficult, if not impossible, for a busy, overworked construction organization to accurately predict how much they can effectively perform and finance. Our industry has yet to accept the reality that every construction operation has very real limits.

There are two critical growth limits that are either unknown, not believed, or regularly ignored. They are the limit on how much work a construction operation can produce at a profit and can finance at a profit. The critical phrase here is *at a profit*. What obscures this limit is the fact that there is no limit, or at least a much higher limit, on how much work a construction operation can produce at a loss. One might argue that a loss would be immediately obvious, but that is not true if the firm is growing and particularly during rapid growth. During rapid and/or continuous expansion, the measurement of profit is almost always artificially exaggerated to the extent that growth can shield financial problems for a long time. The reality that profits are often exaggerated by the common mismanagement of the *work-in-progress* method of accounting has already been addressed.

Experience confirms that during rapid growth, the efficiency of *work-in-progress* field reporting processes is often affected by the staff shortages described above. Getting the work done supersedes reporting on the work, and timeliness pressures result in less attention to detail. This is a formula for potential inaccuracy. As previously described, built-in industry processes and natural tendencies make certain that percent complete is never understated. Thus, profits on work-in-process during rapid growth tend to be overstated. At the same time, actual profits earned may suffer from the increased potential for negative production issues commonly experienced during rapid growth. It is difficult to gear up quickly enough or solidly enough to hold profit margins during growth periods, particularly with inflation, a shrinking labor market, and diminishing skill levels.

Rapid growth puts a strain on a company's key people and systems, and sustained growth doesn't allow for a reasonable training period. Continuous growth doesn't give an organization a chance to test new people or systems before more new people and systems are added. If performance deteriorates as a result of growth, it will only be discovered after the additional volume and people are taken on. It is difficult to develop and instill corrective measures while people and systems are stretched and while overworked managers are coping with *too much work*. Some companies don't recover from this scenario. The question arises, if a business is struggling during market growth, what will happen to it during the inevitable next market slowdown?

Management Constraints

There are fundamental constraints to healthy and sustainable growth. The management of growth requires careful balancing of sales objectives with the firm's operating efficiency and financial resources. The trick is to determine what sales growth rate is consistent with the realities of the company and the marketplace. Each construction organization has very real limits to its abilities, resources, and available capital. Each organization is capable of doing just so much. During periods of rapid growth, closely held construction companies are changed so much that they, in fact, become new organizations. As such, they are also untested in that the new configuration has no proven track record of success. And this is happening while they have an increased amount of work to produce. Most don't realize, and others simply refuse to recognize, that their prior organization that was once so successful is gone forever.

If an organization is to grow, its management must grow, and that growth must be qualitative, not just quantitative. Qualitative organizational growth takes time and needs to occur prior to sales growth. It takes more time to grow management than it takes to capture more work. But most companies decide that they can only increase management after the additional work is on hand—not before. Growth just for the sake of growth is risky in any business, but growing in the construction business is far more complicated than it is given credit for. Growing a business without growing management capabilities defies logic. It is a natural tendency for successful entrepreneurs to *feel* that they can run any size business, but there is no empirical evidence that supports this. It is like saying you can promote a well-educated, promising lieutenant to general, which makes no sense. People advance in life and in rank through education, training, and experience. And all of these take time, and there is an essential sequence to the process. You can get the experience portion of piloting a plane before being educated in the process (ground school). However, we do see young entrepreneurs attempt to grow a business overnight.

Growth Risks

Incremental growth instead of sustained growth may seem unnecessary—even unnatural—but it is the best way to control the inherent risk in growing beyond 15%. Utilizing a series of growth periods, firms can expand, then evaluate, and then expand and evaluate again. This way an organization can reevaluate and recover after a bad test. This approach is recommended in lieu of continuous growth because it's almost impossible to identify, evaluate, and fix problems as you go along. An analogy would be attempting to fix a problem with your plane while it is in flight. Periodic evaluation is a prudent approach to controlling growth risk. During continuous growth a company outgrows its people and systems so often that you never really have the same organization long enough to truly evaluate it. Such a company is at constant risk with an ever-changing team. In some cases, fatigue or failure is just a matter of time.

The common practice of capturing more work before expanding existing employees and equipment resources results in increased cash flow. The reason is that the existing employees and equipment tend to get *overworked* accomplishing the increased workload before the expanded resources can be engaged. These *improvised* efforts produce the work at a reduced cost compared to the expanded employee and equipment costs anticipated in the estimate. This improves cash flow; however, during continuous growth, this process causes *burnout* of both employees and equipment. Profits are also impacted by the cost of overtime premiums.

Rapid growth occurs almost exclusively during a growth market, and a growth market, by definition, means full employment for the trades. A growth market presents contractors with the opportunity to capture more work at a time when there is minimal opportunity to hire additional qualified employees. In a growth market, most qualified employees are already employed. A growth market restricts the opportunity to purchase or rent more equipment at pre-existing prices because equipment is in short supply. Therefore, the common practice in our industry of capturing more work in a growing market before engaging the necessary resources *never* works as planned. These realities are confirmed by extensive research, and while I am not suggesting readers engage in research, I am suggesting we learn from it. A growing market presents opportunities for growth, but the opportunity can't alter the capacity limits that each construction company has. These limits are also unaffected by a reader's belief that said limits don't exist.

Financial Limits

In construction, *Growth Eats Cash*. One of the main reasons is because contractors put the work in place and then wait for their money. A cursory look at other industries indicates that this is not the norm. In most industries vendors are paid in advance—in part or in full. They always have secured credit in place or require some form of payment assurance before providing anything. Contrast this with the construction industry, where profitable companies have failed by running out of cash and credit while waiting for payments. We have already established that there are limits to how much work an organization can put in place at a profit. There are also limits to how much profitable work an organization can finance. These two limits are independent of each other, and the fact that neither is widely known nor believed changes nothing.

It has been clearly demonstrated, and most readers already understand that performance issues can and have caused construction business failures, either independently or combined with financial issues. What should not surprise readers is that financial issues alone can and have caused construction business failures. A review of a couple of case studies will be informative.

Case 1: A well-run mid-size building contractor was converting from one accounting software program to another. During the installation of the program and transfer of data from the old system, something went wrong. Financial data were lost or compromised. Experts were brought in, and after months of effort, the situation was still not rectified. In the meantime, the end of the contractor's fiscal year had come and gone, and the outside accounting firm was unable to complete the firm's annual financial statements. Months after the statements were overdue, the surety became nervous because 100% of the firm's work was bonded. The firm continued to need bonds for new work, and the surety sent my firm in to investigate. All of the major projects looked OK compared to the estimates, but there were no current financial statements or other accounting information available to verify income, expenses, and profitability. Because the firm desperately needed new work, they agreed to certain financial restrictions and access required by their surety.

My firm's director of forensic accounting reviewed the missing data issue and reported that some of the lost data may never be retrieved. We decided the only way to produce the critically needed financial statements was to go back to where the old accounting software was abandoned and bring that accounting information system up to date manually. We were able to convince the contractor to pull his accounting staff off trying to update the new system to assist in our effort. It took more than six weeks of working long hours, seven days a week, to produce a current financial statement. We were able to report to the surety that the contractor's business had continued to be as profitable to date as it had been previously. As a result, their bonding capacity was restored. The lesson is clear. Putting construction in place for a profit is only one of the necessary elements for success in the construction business. An equally important, and some might suggest more important, element, is managing the accounting and financial records. Accounting and financial calculations and record keeping are not *ancillary* to business. They are an integral part of business.

Case 2: A very large highway contractor wanted to bid on a project that was exactly their type of work, but the size of the project was much larger than any project they had ever done. The project was the construction of a many-mile-long two-lane road from a rural highway to a mining site in the middle of nowhere. The contract would be with a valued private client they had performed a lot of work for. While the total bond amount would not exceed their overall bond limit, the surety was reluctant to issue the bond because the contract amount was way above their single project bond limit. The surety recommended that the contractor call my firm for advice. I told the client that I strongly recommend against taking on a single project that is so much larger than anything they had ever attempted. They said that they were familiar with my position, which is why they want all aspects of the project evaluated so that they can fully understand any and all risks associated with it. It was clear they wanted to do the job.

After reviewing the design documents, the client's financial strength, and visiting the site, my position had not changed. It was clear that the work itself was no challenge for the client's organization, but this was a financial exposure that I could not recommend. The job would take more than a year, and I demonstrated to the client that their monthly cash outlay would be huge. That combined with retainage would require borrowing by the third month into the project, and before 70% completion, the debt would exceed their current borrowing capacity. They would not be able to finish the project without a substantial increase in their borrowing capacity or some outside source of additional funding. In spite of my reluctance, they still wanted the job and wanted me to explore all options.

When I met with the mining company, who fortunately favored my client, my first suggestion was the elimination of retainage. This was not possible because the mining company's lenders would not allow it. My next option was to break the project into three separate contracts, which were not complicated because the project was linear. Each contract could cover one-third of the many miles of road. With the reduced amount of each contract and the earlier collection of retainage withheld, my client would then be able to manage the cash flow. The contract language remained unchanged. Just the number of each of the three contracts was different. This was a financial solution because it was a financial problem, not a performance problem. My position on the size of the project remains unchanged. I trust this further demonstrates the distinction between performance issues and financial issues.

Case 3: A rapidly growing and quite profitable mid-size trade contractor passed away suddenly and unexpectedly. The contractor personally ran all aspects of the company, and no managers offered any ideas about how to proceed, so the firm was temporarily closed until after the funeral. By that point the survivor's attorney realized that this first-generation company had plenty of highly qualified middle managers, but there were no senior-level managers anywhere near qualified to run the business. All the firm's contracts were bonded, so on behalf of the survivor, the attorney advised the surety company of the situation and requested help.

The surety had no choice but to complete the projects. It took almost two years to finish all the projects in progress. The contracts were so well priced, and the field management and crews so well organized, that all were completed close to original budgets. When the punch lists were taken care of, final payments collected, and all bills, including miscellaneous expenses, paid, there was a reasonable amount left over. The leftover funds (which under different circumstances would have been referred to as profit) belonged to the survivor. This firm had ceased to exist suddenly, and all concerned anticipated a massive financial loss. A loss that never materialized because of the firm's prior financial health. The completion of the work certainly cost more than it would have if it had been completed by the very competent contractor, but the expected loss was prevented by prior superior financial management.

Unanticipated Results

There is an occasional, unusual, and unanticipated result of growth that occurs, primarily in small, closely held companies. Because it happens less regularly than the universal growth issues mentioned earlier, this may be of limited interest to some readers. When people work together as a group for some time to achieve common objectives, it tends to become a group effort, which adds satisfaction to the work process. When a firm grows, new people are introduced to the group, and it disrupts the group dynamic, and work becomes work again. It is not a very big deal, but a few small contractors have lamented the fact of having outgrown their organizations. More than one said that *it took the fun out of the work.*

For example: An entrepreneur friend called to tell me he had just made plans to triple the size of his offices because he had just landed several large clients who wanted him to do all their work. He had secured a lease on half a floor of a multi-story office building. I congratulated him, and we discussed some of the details when he said, *you don't sound excited about this*. I deleted it, but he insisted I explain to him what I was thinking, which I did. A little background. My friend had the same employees for many years, who were all squeezed into a small office space in a converted house. I knew his firm well, and they were a very tightly knit group that worked really well together.

Several times a year my friend would surprise this group of hard-working employees with a group celebration. For example, he would tell everyone there would be a group lunch on Friday, and then at 11 AM on Friday two limousines would appear at the office to take all of them hot-air ballooning, or to a major league baseball game, or to a concert, a play, or the finest restaurant in town. They had their own softball team that played against other small businesses in town. The office Christmas party included wives and children and was at my friend's home, where he also had company barbecues several times a year. I was occasionally invited to a softball game or one of their events and greatly admired the camaraderie and team spirit he fostered. It was a fun place to work, and his profit percentage was off the charts. His employee bonuses also reflected the profits.

What I was forced to explain was that my experience was that when groups like his are expanded by more than one person at a time, the camaraderie is always affected and usually lost. He actually appreciated my telling him that and said he would not let that happen. I cautioned him that it was a natural reaction and one not easily prevented. They moved into a new office shortly thereafter, and within six months he tripled his staff, and within a year it doubled again. The next time I visited I had to go through a receptionist before his administrative assistant grilled me about what I wanted and didn't believe I was just a friend. My friend, of course, knew nothing about that because his private office was so far removed from the front, he had no line of sight on anyone. He explained that things were going better than expected, they were still hiring, and he was trying to get more space in the building. We were both very happy. About 18 months later my friend called me to say I had jinxed him by saying people would leave. He said the second of his old group had just quit. Six months later, three more left, and only three of the original 10 remained after two years.

Sometime after that I asked my friend if he would have done things differently, and he said he was not sure, but that he sure missed his early days in business and that he was not enjoying his new position as much. He said he hardly ever got to the field anymore, did not like being stuck in the office all day, and that all he did was solve personnel and client problems. He said, all of a sudden, he works alone and that being the boss is no fun. 15 years later he sold his business, and he told me it was to save his sanity and his marriage. He also said he wished he had never grown his business as he did because his first 12 years in business were the best in his life. He shared that he had been so busy he did not even notice that his kids had grown up, and he was not there to enjoy that.

Note: These last comments were included because there is a lesson to be learned about the personal impact of growing an organization on the business owners and senior managers.

Summary

The definition of *growth* of a construction enterprise was redefined as *profitable growth* as opposed to growth in size, because *unprofitable growth* in a construction business context is meaningless. Anyone can increase sales by capturing a lot of work at a loss. This chapter was devoted to demonstrating that thinking a construction company's size is optional is a fallacy. Growth is not a contractor's choice. Profitable growth is a carefully planned activity and is limited by both internal and external factors. Each company has distinct limits that can be known in advance. There are two separate and distinct limiting factors on growth, performance, and financial.

Perhaps the biggest lesson is that exceeding growth limits increases risk and can cause business failure, which makes growth a little less attractive than it may seem. Readers were advised that growth always includes risk and that growth opportunities cannot be evaluated without considering the risks associated with the growth activity. These facts are reinforced when you consider that most of the common elements of construction business failure are driven by growth. There is nothing wrong with growth as long as it is profitable growth.

11 Profit Trumps Volume (Managerial Maturity)

Introduction

The construction industry is volume-driven. Most construction professionals agree, and the few that claim not to agree generally say that profit *also matters*. My experience is that most contractors exclusively pursue growth in total sales until they are satisfied with the size of their firm—at which point they may begin to focus some energy on profit. There is certainly tacit agreement, demonstrated by contractor behavior, that growth in sales and size of company is a primary objective.

> Note: Tacit agreement is defined as: an agreement that is understood through action conduct or silence, where mutual consent is implied by the circumstances or behaviors of the parties involved.

In spite of the fact that it is common industry knowledge that small contractors make a greater percentage of profit than mid-size contractors, who make a greater percentage than large contractors, nobody seems to want to remain a small contractor. This is also in spite of the fact that everyone knows that small contractors are at less risk than mid-size contractors.

This does not imply that everyone thinks profits are not important. It means that industry practices confirm the industry believes that size matters much more than percentage of profit or business risk. Business decisions are based on advancing goals that are sometimes incompatible while others are mutually exclusive, which means hard choices have to be made. For example, pricing the work has an impact on profit and growth. A lower price increases the probability of getting the work (growth), while a higher price increases the probability of greater profit. Choices are also impacted by competing goals and motivations. When asked, many contractors claim to equate profit and growth equally but, in practice, consciously and deliberately choose growth over profit every time.

Risk Matters

Business decisions and choices should never be evaluated without considering the amount of risk involved. However, researching the survey responses of contractors and construction professionals reveals that risk was not considered in the majority of management's decisions. It is long past the time we need to introduce more science into our industry's decision making. Some basics about risk were introduced in earlier chapters. In this chapter, we explore the risks associated with prioritizing growth over profit. It helps us to understand that profit and growth risks are in separate categories. Decisions involving or affecting profit are *financial risks*, while decisions involving or affecting growth (in volume or total sales) are *operational risks*. These categories are closely

DOI: 10.1201/9781003500315-12

related, like two sides of a coin, and you can't have one without the other, so determining if one is more important than the other can be confusing. That is what this chapter is about.

There are times when decisions about profit or growth may be mutually exclusive to the extent that you can have one or the other, but not both. To deal with the choices, it will be necessary to explore motivations that impact decision making. For example, it may not be obvious to a contractor when conflicting motivations are involved when pricing the work. Pricing the work low increases the potential of capturing the work, so you might say that strong motivation to get the work trumps profit. But if profit is the primary goal, that pushes growth down on the priorities list. Like so much in business, the correct choice is not always obvious.

Growth Simulation Exercise

> Note: This simulation is not limited to residential construction. A residential example was used for simplicity and because there is a buyer for the work produced (the house). Being for sale immediately built provides the short production cycle needed for the simulation. The business process is the lesson, not the product.

This learning tool was developed by the author to teach construction risk and pricing principles to graduate students and construction professionals pursuing a master's degree in construction management (Figure 11.1). The purpose was to demonstrate that there are very real financial and operational capacity limits in the construction business and to substantiate how motivation influences decisions. The simulation tool was similar to a board game like Monopoly, with dice, paper money,

Figure 11.1 A classroom growth simulation illustrates how the drive for volume often overshadows profit.

and little plastic houses with properties to place them on. Teams of three students would form construction companies, and one student each would become the material supplier, the banker, and the real estate broker. Each construction company started with the same amount of money, which they would use to purchase materials from their classmate, who would play the material supplier. The materials came in the form of a completed house, which was a little plastic Monopoly house. The contractor students placed the houses on various properties they bought from their classmate, who would play the real estate agent. The contractor team would then sell the houses to the banker for a profit.

Each company started with enough money to buy the property and material to build three houses and to cover the overhead costs for the three months it would take to build the three houses. Each construction company also started with a bank line of credit sufficient to build another three houses, including the property. The interest rate on the line of credit would be the prime rate plus one point (one point equals 1%). Each company would also have a separate agreement from the bank for secured lending, which was guaranteed by both the assets of the company and the partners personally. The banker required each member of the contractor team to sign a corporate loan guarantee for the corporation and separate corporate and personal loan guarantees for the secured loan. The secure lending limit was enough to buy properties and materials to build up to six houses. The interest rate for this lending would be three points over prime. The students knew all the rules well in advance of the three-hour class where the learning experience would occur. The simulation would consist of six to ten separate rounds or cycles of buying property and materials, building the houses, and selling the houses. Each round would take 10–15 minutes, in which all decisions, purchases, and sales would be made for that cycle.

The students were made aware that in the real business world there is always some business risk, which needed to be added to this simulation. The participants learned that the uncertainties contractors have include:

- Not knowing for certain the availability of properties that will be required for development.
- Uncertainty about the availability and quality of tradespeople and subcontractors that will be needed.
- Lack of information about the number, timing, and qualifications of potential home buyers.

Simulating Uncertainties

To simulate the risks, contractor teams will be required to build without knowing the housing market. They will need to decide how many houses to construct in each round without knowing how many houses would be purchased after they are built. The number of houses to be purchased would be announced after the teams had purchased the properties and materials and built the number of houses they had decided to build. The students have already learned that statistically as contractors grow, their profit margins diminish—the reason being that the advantages of having management's direct attention to the work and the lower overhead of small contractors will disappear with size. To simulate this reality, profit per house will decrease by one half of one percent after five houses and at five-house intervals. Overhead as a percentage is included in the profit in this simulation and does not reduce with size. The house sale price remains consistent through the exercise.

The number of houses to be purchased by the banker will be listed on a stack of three-by-five cards that will be turned over to reveal the number near the end of each round. At that time the teams will be able to calculate their profit for the round. They will also have to pay the interest on any outstanding debt. Outstanding debt will, of course, count against their debt limit for borrowing capacity at the beginning of the next round.

> Note: If this seems confusing, it is a simulation of exactly the type of decisions and choices that construction professionals make every day. It should be noted that many times business decisions are made linearly, or as they come up. It may be wise to evaluate all available information collectively and attempt to project long-term impacts of the decision being considered. The students in question are adults employed full-time in the construction industry and attend graduate school at night in a master's degree program. They are a reasonable cross section of entry- or mid-level construction professionals, albeit with limited experience. Few would have been in positions where they make decisions of this nature.

The participants have studied the simulation process as a homework assignment and are again briefed on the rules of the game at the start of class. They are reminded that the construction market they will be working in will be a growing market. They receive advanced information that the number of houses to be purchased in round one will be three, and that in all subsequent rounds, the number of houses to be purchased will always be greater than the prior round. They are also informed that the prime interest rate is 10%. (The program was developed when the prime was high.) The teams are then given time to discuss their initial strategy, and they can ask questions before the start of round one. The students already know the material costs and sales price of a house and that the combined overhead and profit amount is 33.33% (for simplicity).

> Note: What experienced construction professionals should be able to figure out at this point, if they take the time, is that if any team builds three houses with the money on hand, they will earn enough profit to build four houses in the second round and five and six houses in third and fourth rounds without borrowing. By round five they can build eight houses with just the money on hand when they started. This is the maximum profit percentage that can be earned over time with no added risk because the firm would pay no interest. It is also at minimum risk because the firm takes on no debt. The students are not told this, but they should be capable of figuring it out because these types of information and strategies have been presented in previous classes.

Simulating the Construction Process

In round one, occasionally, a team will build only three houses, but most borrow money to build up to six houses. Occasionally, a team will borrow on the secured loan and build 12 houses in the first round. There is always a lot of excitement waiting for the announcement of how many houses will be purchased. The first card of a stack of three-by-five cards in the front of the room with the number of how many houses will be purchased in each round is turned over. Five houses will be purchased. Lots of work for the bankers, including interest payment calculations for those who built more than five houses. Those who built three houses do fine, and those who built six are overjoyed. Now the strategizing begins anew, and the discussions are almost exclusively about what the next number will be. Everyone knows that it will definitely be higher than five, and decisions seem to center around trying to figure out the maximum they should build.

Round two begins, and by this point, everyone is seriously into it as if it were real money and was their own money. No one ever builds less than six houses in round two, and the hardest decisions are for teams that already have unsold houses in inventory from round one. Decisions are made and everyone waits. The next purchase number is seven for round two. In this round, just about every team had built between six and twelve houses. Lots of banking activity goes on, and more houses are already being built for the next round.

In round three, the number of houses that will be purchased is announced as eight. There is great disappointment because everyone was sure it would be at least nine. Some say, *what's going on here, you broke the pattern.* Of course, two numbers do not constitute a pattern, but because there was no risk that the market could get smaller, everyone was laser-focused on the growth.

By round four, discussions are almost exclusively about what the next number will be, with much talk about making up for the round three disappointment. Without recognizing it, the entire class is obsessed with volume—with total sales. No one is talking about *profits, past performance,* or *inventory*. You could have heard a pin drop when the purchase number was about to be turned over. Everyone has been fixated on sales as if this were a contest and the winner will be the one who builds the most houses. Of course, this is NOT a contest. This is and was announced as a *business* simulation. The bankers, material suppliers, and real estate agents are doing well in their roles, while the contractor teams all seem to be exclusively focused on growth. There are typically five to eight groups of a banker, supplier, agent, and contractor team that work independently with little contact. The same thing happened with different participants every time this simulation was run (almost like mass hysteria).

Lessons Learned from the Simulation

The number of houses to be purchased in round four is announced as ten. This drives a lot of discussion, particularly with groups with growing inventory that continue to pay interest to the banks. What these student contractors are beginning to realize is that their inventory will be profitable when sold, but their cash flow is tightening up. They are learning the distinction between profit and cash flow. Some borrowed money on their line of credit just because of the need for cash flow. Others who had maxed out their line of credit attempt to increase borrowing on the secured note only to find out that they have no collateral. The bank already has lien rights on their unsold houses, so some contractors were approaching their debt limit. Some were already at their debt limit with no further credit available. All of the groups are learning something about *borrowing and lending policies* that they did not previously understand.

During round four, there is usually at least one team that has run out of cash and can't pay their interest payments on unsold houses. They usually try to convince the banker to lend them more money on their secured credit line which, of course, the bank cannot do, and the first insolvency (or more) occurs. At this point the student banker usually wants to increase the borrowing, and the class is paused long enough to hear a lesson in credit granting. The lesson comes directly from the fine print in the simulation directions that each participant was supposed to study as homework the previous week.

This business lesson stresses how it is an absolute necessity to read all of the fine print before signing anything for your business or personally, even if you believe you know what it says. And if you don't understand it, hire someone who does to explain it to you. The lesson is that even though it is common after years of signing similar documents, people in general and contractors in particular tend to stop reading the fine print. It is still a dangerous mistake. This is because the fine print on what appear to be similar documents often changes over time as regulations, laws, and practices evolve. Students are warned that while they may hear a contractor say, *I don't have time to read all this junk*, or, *my job is to build things, not to read documents*, these statements are incorrect. Nothing excuses our professional responsibility to understand the agreements we execute. The students are advised that they may also hear contractors say at some point: *If I had known what that document said (or meant), I never would have signed it.* Accordingly, the following statement bears repeating:

If we are ever going to reduce the failure rate
in our industry, we need to change behaviors.

In this simulation exercise, few students carefully read the loan documents that they sign, and some who read them don't fully understand them. The class instructions advised the students that if there was anything they did not understand, they should ask the instructor for clarification. Few students ask for clarification. It is pointed out to them in the debrief that many businesspeople make the mistake of not asking questions because they think they are supposed to know the answers or are embarrassed to ask. The question is posed: *Should a businessperson try to look smart or do whatever it takes to make informed decisions*? The fine print on the notes the participants signed explains the terms and lien rights. The students are learning that some of their unsold houses can't be used as collateral for a second loan because they are already collateral for the secured loan that provided the funds to build them.

Moving on to round five, the purchase is announced at ten houses. This seems to convince everyone that the advancing *pattern* (that never was a pattern) is back. Round six is 13 houses, and the group now seems sure that the number of houses purchased will continue to grow, even more rapidly. All subsequent rounds increase by only one house, and by then most groups have over-invested and are paying interest on houses built several rounds prior. Depending on the number of insolvencies, the simulation ends at rounds six, seven, or eight with plenty of time left in the three-hour class for teaching and discussion. The teaching includes what was learned and what decisions might have been made differently.

Business vs Building

An important point in this *construction* simulation is that no building or construction tools were used. That is because construction tradespeople and field managers *build* the work, and actually building these houses is only a portion of the business cycle. A larger portion of the earning process is *managing* the business and financial portion of the cycle. It was exclusively management decisions that caused all these *simulated* insolvencies and business failures. This is ALWAYS an *ah-ha moment* for the student participants. Hopefully it is also for readers because the prevailing industry-wide belief is that the construction industry is about building things. Construction is a BUSINESS, and producing the built environment is its product. More specifically, construction is a for-profit business. Construction is more like a monopoly game than an erector set.

CONSTRUCTION IS A BUSINESS

It was critical for the participants in the simulation to fully understand that businesses maximize profits, not growth. There is nothing wrong with growth as long as it is not the primary objective. Growth should never be at the expense of profit because only *profitable growth* makes any sense. Every team was concentrating on growth from the very beginning, and this was a huge learning opportunity and was discussed at length. The students saw that clearly after the fact and readily admitted that it happened. What they had no explanation for was why it happened. During discussions, one or more students would always remember that they were taught the profit-first principle in class but still could not explain why the drive for growth took over so quickly. This would always lead to a long discussion with some students quoting things they learned in psychology or philosophy classes about human nature, etc. They would eventually recognize the issues and learn that business principles must supersede personal urges. At most sessions it would come out that one or more members of their team would have discussed profit issues, but that somehow growth always seemed more important. A solid lesson because growth is so attractive and apparently a great motivator.

There would always be the question of how a team could have any way of knowing how many houses to attempt initially. Participants were reminded that both financial and operational capacity concepts were presented in prior classes. Some students would immediately exclaim that we could have used those concepts to figure out how to optimize return on investment and effort. This was always a big breakthrough that would lead to the question, did anyone consider starting out with some business planning or setting business goals? Few did, and it is always brought up that such planning requires experience, and all were relatively young and inexperienced. This opens the door to a meaningful discussion about seeking advice from people with more experience and that age has less to do with it than what a person has studied and experienced. This usually leads to stories from students who made poor decisions and later realized they could have asked someone older and more experienced for advice and avoided the wrong decisions. Another great lesson.

Borrowing

Students were always asked why they borrowed so much so quickly. And the response was almost always the same—that it was so easy. Some blamed it on the rules, saying borrowing should not have been made so easy. This, of course, is a perfect invitation to explain that borrowing corporately and personally is in fact very easy. And to explain that easy borrowing is why so many companies and people get into debt they can't handle. Students learn that lending is based on a person's or company's credit rating and that those with a good credit rating get a lower interest rate than those with poor credit, but everyone can borrow. Borrowing can be a trap or a benefit, but it is a science. A simple rule of thumb is to borrow to *make* money, not to *have* money. Students are warned that they hear: borrow what you need—not what you want, but that does not work. Borrow only what you are absolutely certain you can pay back—or do without. Borrowing is more of a *trap* than a *benefit*. Some will disagree, but you might want to use this rule of thumb: If you have to sign personally, does it mean that the lender does not think you can pay it back? If so, maybe you can't pay it back.

Risk

The class is advised that the subject of *risk* did not seem to come up much in their discussions. Most said that they did not see where there was much risk in the simulation. A lesson about risk fits in well. The *recognition* of risk is a huge problem because most people misunderstand risk or underestimate it and, as a result, have little concern about it. Risk is generally defined as the potential for harm, danger, or a negative event. However, *business risk* may be closer to *uncertainty,* which is the potential for an outcome to be different from what is expected or planned. Risk may be the single most important topic of the entire simulation because risk management is critical in business. The participants, like all businesspeople, should have discussed the likelihood of the various possible outcomes of each decision they made.

For example, all of the teams spent considerable time speculating on what the first number of homes to be purchased would be. The decision to build three houses should have been easy because the sale of three houses was a certainty. Any more than three was an unknown. By definition, business principles prefer certainty, partly because certainty is encountered so seldom. In addition, building more than three houses would require borrowing and paying interest, which reduces profits. The earnings were certain and therefore risk-free. If more than three were built and not sold, there would be an added cost or penalty of paying additional interest until sold. With no other information to draw on, good business sense dictates, and some may say it demands, selecting the sure thing. After the first round, the teams did not learn anything new, so the only sure thing was the purchase of four houses, so the same logic applies. Anyone who used that logic on the first round should almost automatically build four houses in the second round, but some of those teams built more than four houses. The motivation to speculate on more than four houses could only have been a desire to *grow* the business, which is selecting potential growth over certain risk-free profits.

When round two turned out to be the purchase of five houses, the teams that built four houses had regrets as if they *lost a bet*. The teams that built more than four houses, even five, six or more houses, were ecstatic as if they *won a bet*. The lesson here should be that business is NOT a *game of chance*. In business, if there is an opportunity to avoid risk, take it. And where risk cannot be avoided, try to minimize it. Some will argue that to make money you need to take risks with the motto *no pain no gain*. This is not a business expression but a gambling expression. What the majority of teams neglected to consider was that there is risk in leveraging assets, borrowing money, or incurring debt. The participants failed to recognize that an unknown or unclear market presents a business risk and therefore risk evaluation must be considered in any decision making concerning that market. At the very least, the teams should have recognized and considered the cost of holding onto unsold houses.

It seems that all of the teams understood the value proposition that the houses could be sold at a 33% markup, which after paying their overhead costs, the remainder would be profit. What they did not seem to realize was that the profit on any house built with borrowed money and not immediately sold would be reduced by the ongoing monthly interest cost on the loan. With interest at 10% annually, the monthly cost would be almost 1% a month. In addition, in this simulation the opportunity to sell occurred every three months. Some may argue that a three-month delay in sales is unrealistic, but these master's students were graduate-level engineers, so understanding the math in this *simulation* should have been simple for them. Much like the fine print explained above, following instruction as in a specification or a loan agreement is fundamental to standard business practices. These participants need to stop thinking of themselves as *builders* and realize they are *businesspersons and managers*. An important lesson in this simulation is understanding that *building* is conducted in the field and *business* is conducted in the office.

Builder or Businessperson

The industry's disastrous rate of business failures is not a result of an inability to *build* the work. The failure rate is a result of an inability to manage finance, accounting, economic, and business issues. Those unable to build the work for a profit, or those who are unwilling to work long hours, fail during the first ten years in business. Those able to build for a profit continue indefinitely or until a non-construction, non-building issue goes wrong on the financial/operational side of the business (Figure 11.2). This is an industry reality that is statistically confirmed by research and should be obvious in light of the high failure rate, which is difficult for successful construction professionals to accept. They argue: *We have been in business all these years or for generations and have not encountered what you are talking about*. Or another popular response is: *Sure, my third-generation competitor went out of business but it's because they didn't know what they were doing.*

A first step in process improvement is recognizing the need. Those who believe our industry or their businesses are doing OK may be interested to learn that we have some unrecognized shortcomings. As mentioned earlier, a 30-year study of contractor financial performance demonstrates

Figure 11.2 Managerial maturity means prioritizing profit over sheer volume, reducing risk, and ensuring long-term survival.

that the accounting systems we use, impacted by how we use them, regularly and continually overstate our profitability. This is embarrassing, and nobody wants to hear it, but unless we face reality, we will never recognize or agree on the need for some improvements in processes. An analysis of the published financial data of thousands of mid-size and large construction companies indicated profit overstatements on average of almost 1%.

That does not seem like much until you consider that the average published profits of these thousands of firms average less than 5%. And there were no firms in the published data that were losing money, and we all know by the failure rate that such firms existed during that period but simply did not offer their data for publication. What does this mean? It means that readers comfortable with their current and historic performance should ask themselves if they would remain satisfied if their certified financial statement profits were 1% less. Let's not forget if the average profits were overstated by 1%, then there were some profits overstated by more and some overstated by less. Let's all be honest with ourselves and admit that we have experienced the situation where we reported a project percentage complete for work in progress that later deteriorated. The question is, did earned profits deteriorate, or were they inadvertently overstated?

The potential for error is unintentionally built into the percentage of completion method. And there is no possible motive for those reporting the percentages to understate them. However, there are very real practical motives and pressures to overstate them. I am not critical because I fully understand how our industry got here. I am simply advocating for a realistic evaluation of the need for performance and profit improvement in our industry. We are doing an amazing job producing the built environment. We just need to be paid better to do it, and that is up to us, not others.

Summary

As stated so many times already, the purpose of this book is to attempt to tell what may be the greatest industry in the world that we can do better. I am not saying we are doing it wrong because I know we are doing it exactly as we were taught, and we have a proud legacy. But we also have a history of being taken advantage of while both the owners and designers we worked for kept altering the language in the contracts that govern the conditions of our relationships. As the authors of those documents, the architects and owners have, over decades, taken complete *authority* over the construction process. At the same time placing on us, the contractor, complete *responsibility* for the constructed results. This has eroded our profits to the breaking point and expanded our risk to the point that small and mid-size contractors are being forced out of business. Unless we recognize these realities, our industry will continue to consolidate until only huge contractors dominate, which has already happened in many areas of the world.

While we can't change the industry in the short term, we must begin to recognize the ineffective legacy business practices that constrain our profitability and performance. Until we agree on the constraints, we can't begin to do anything about them, which is the problem. These legacy constraints came upon our industry so slowly, over decades, beginning before some readers were born, so we take them for granted. Most accept them as *natural*, believing that *if that's the way it has always been, that's the way it will always be*. Contributing to our industry's acceptance is that each of these constraints is subtle, even elusive, and seems *small* in the scheme of things. And each individual constraint may seem harmless if compared to bigger and more obvious industry issues we deal with. The problem is it's a comparison of apples and oranges. The big constraints, like being outdoors, the weather, remote or congested sites, each project being different, and the like, are built into and a part of the industry, where nothing can be done about them. Nor should anything be done about them. They are why the world needs our expertise. It's what we do.

Every chapter in this book has focused on *constraints* that undermine profitability and performance—constraints that we actually can do something about. By definition, this means we, as construction professionals, are often accepting these limitations without even realizing it. The real issue is that many of us no longer recognize them as constraints at all; instead, we treat them as if they are as unchangeable as the weather. But unlike weather, these business practices and assumptions are not beyond our control. They only remain in place because we allow them to. Recognizing them for what they are is the first step to changing them—and to reducing the unacceptably high rate of business failures that continues to plague our industry.

12 Management Succession Is Treacherous

Introduction

A closely held company is defined at any given point in its development as the sum and substance of its experience. The majority of the sum and substance is created by and resides in the company's people. More specifically, it's created by and resides in the organization's key managers. Some of the sum and substance becomes institutionalized in the company, such as attitudes towards customers, work ethics, values, and other qualities. These qualities are woven into the *fabric* of the organization by the example, training, and mentoring of key managers. Some of the firm's sum and substance can be transferred from one person to another with deliberate and diligent effort. Processes like customer relationships, union contacts, and production and process knowledge are transferable. However, a considerable portion of a firm's sum and substance cannot be institutionalized, is not transferable, and, despite a company's best efforts, is forever lost to the organization when a key person leaves.

Things like insights into business or project risks, the development of innovative processes, actions born out of years of company-specific experience, masterful hiring selections, solutions to personnel issues, decision making based on an ability to read between the lines, being able to *hear* what is not being said, and much more. These innate personal talents are learned and honed over decades and are lost to the company when a key manager departs. This chapter is about the steps necessary to minimize the loss of critical portions of a firm's sum and substance at succession. The first step is recognizing the risks that arise during and after management succession. The next is acknowledging the need to prepare for it, followed by the most straightforward step of all—developing an appropriate succession plan.

Key Managers

The incidence of business distress and failure of construction firms in the United States increases exponentially during a period of five to ten years after leadership succession. When the contractor, chief executive officer, or key top managers leave the firm for any reason, the organization is never the same again. It is permanently changed and has no track record or experience under new management. Some will argue that after succession the firm is the same organization, with one less person, but it is absolutely not that simple. Identification of those who are key to success in activities and processes within the profit chain takes some introspection because we are not known as an industry that analyzes our activities.

There are very serious succession risks in *all* small and mid-size construction firms. Some large firms have formal and effective succession strategies, but certainly not all. Succession risks are minimized by the *quality* of the succession plan. Succession in most small firms

DOI: 10.1201/9781003500315-13

includes one key person, the contractor, who is often the founder. Partners active in the business may be included. Mid-size firms often have key managers that may not be owners but possess indispensable and sometimes irreplaceable knowledge, talents, or information. It is critical that top management understands who is essential to the success of any portion of their operation and, more specifically, who possesses any of the sum and substance of the organization. In large companies there are always portions of the firm's sum and substance residing in multiple key managers.

Why Plan for Succession?

Management succession is not an everyday occurrence but is a critical business issue by virtue of the fact that it is included in one of the five common causes of construction business failure under *managerial maturity* (Figure 12.1). Succession in this context includes any managers that are critical to the success of an organization as well as the more obvious founders, contractors, and CEOs. When a key manager departs unexpectedly, firms scramble for a replacement, incur losses, and always insist that there was nothing they could have done about it. While succession may not occur very frequently, it can and does occur unexpectedly. And when it is a surprise, it is more lethal than when anticipated. Even when anticipated, experience confirms that in construction, it is seldom planned for in detail. It is apparent that in this industry, more effort is put into planning

Figure 12.1 Succession planning is critical to minimize the risks of losing knowledge and leadership when key managers depart.

retirement parties than succession planning. This is one of the reasons successions cause so many business failures.

For every firm, it is an eventuality that a critically important manager or founder will retire or leave. It's not a matter of if, it's a matter of when. Therefore, planning for the succession is appropriate and necessary, leaving only the question of when to establish a plan. Succession planning is a one-time event and may require updates as time goes on. The plan takes hours or days to compile, and it should cover three to four years prior to actual succession, depending on the position and the candidate. Therefore, the planning takes place a minimum of three to four years prior to the anticipated retirement date. However, this does not address unexpected departures, and some will say *so what?* Others will say *now you want us to plan for the unexpected.* My response is yes. If you have fire extinguishers or first aid kits on your projects and in your offices, you are planning for the unexpected. In this case, the planning will take place later anyway. Preparing in advance will not cost any more, but it does provide impeccable risk control which saves time and money should the unexpected occur. The exceptional contractor has these plans on their shelf ready to go. The ideal time to plan is when any manager is promoted to a position that is essential within the profit chain or to the success of the operation. It is simply high-quality risk management.

Unfortunately, there is often no notice when a high-level key person leaves, as it is usually unexpected. For key senior management departures, a minimum of four years of preparation is appropriate. When there is uncertainty about when a person may leave, such as sometime within the next five years; or an announced intention to retire at a non-specific length of time, like two to three years, or sometime after five years, or depending on health issues, it is prudent to plan for the shortest time possible, and if the person stays longer, all the better. The more senior the person or the greater the amount of sum and substance presumed to be lost, the greater the effort and resources that should be applied to the succession planning.

Succession Planning Process

A complicated undertaking is simplified by breaking it down into a step-by-step process. The selection of who should develop this plan is complicated for small firms because of the limited number of managers, but for a smaller company there are fewer people, and a basic, uncomplicated plan outline may suffice. For any company of any significant size, the selection should be any manager that is a capable planner and willing to seek help from others and from published information. The amount of detail is optional, and an outline is usually a sufficient guide to implementation. These are the basic steps which will be addressed in detail:

- Discover, evaluate, and delineate exactly what sum and substance (knowledge/information/talent) is captured in a key person well before they leave the organization.
- Determine which, if any, of the substances are already, or can, be, institutionalized.
- Identify the successor and how the proprietary knowledge, processes, etc. will be transferred to them—by who and by when. (The transfer process can and usually does take years.)
- Determine which substances are not transferable and will be lost to the organization. Attempt to define the impact that this loss of substance may have on future processes or decisions. Determine if there is some way to compensate for the impact on future decisions, such as collaborating with others before making the decisions in question.
- Design a multi-year *Professional Development Plan* (PDP) to strengthen the talents of the replacement manager. (This PDP process usually takes years.)

Discover-Evaluate-Delineate

Just identifying what specialized or proprietary knowledge a key manager planning to retire possesses that others in the company do not is not easy. By definition, this *knowledge* is unknown to others and may not even be recognized by the manager themself because the knowledge is *taken for granted*. The easiest way to explain this is by example: A senior VP of Operations for a very large, highly specialized union firm that built various types of ball fields nationwide was planning to retire, and my firm was engaged to assist in selecting his replacement. The firm had no job description for his position, so the VP was asked to describe his activities. In addition to scheduling the projects and assigning the field management, crews, and company-owned equipment to the jobs, he supervised all the work in progress. He was a major user of the company plane. He did the site visits for the estimating department, interviewed and hired all subcontractors, and negotiated the job site conditions with local union representatives.

This individual was way past retirement age, had been in this position for decades, and was hired and worked for the last generation of ownership long before the current generation took over. He was what some would call *old school* in that he kept no notes or records of his activities. He was actually the only employee, including the current president who did not turn in a detailed description of his expenses or receipts for cash purchases. This individual had built well over 100 projects around the country, including repeat projects for the same client. He had no listing of individuals at subcontractors, union business agents, or client representatives he had done business with. When asked where these might be found, he referred to the inside of old project file folders where he had written the names and phone numbers of all the people he had done business with (often first name only). These folders were in the individual project files, and any older than a few years were off-site.

This VP was a *one-man band*. He seemed to do the work of two people while he traveled and spent more time away from home than any three managers put together. His people-handling skills were also off the charts in that he interacted with and managed such varied personality types as project owner representatives, architects and designers, union business agents, material and equipment providers, and assorted and different divergent subcontractors. Both who he interacted with and how he achieved such consistent and positive results were a mystery. We assigned a consultant to shadow the VP for two weeks as an assistant. Our finding was that a personality type that would have the necessary demeanor and skill sets to successfully interact with these many different types of individuals does not exist. It would require several different personality types and therefore several different people.

This VP was amazing and would be difficult to replace. Perhaps impossible. He was incredibly flexible and talented in that he *mirrored* the personalities he interacted with. He was friendly and polite while communicating persuasively, able to *lead* people to his point of view. This was a classic case where some of the sum and substance responsible for the success of this very large organization was unrecognized and therefore unknown to anyone—neither top management nor the person who *possessed* the sum and substance. Just determining how the VP operated and cataloging knowledge and information that might be able to be institutionalized, transferred to another, or would be lost would take some time. The VP leaving was old and old school but loyal and willing to help, to the extent that he extended his retirement date and would train his replacement. He assisted in selecting a manager in his department who he believed could grow into the job. The process included a number of night courses, a mountain of selected self-study, and years of working alongside the VP as his anointed replacement. It worked.

The defense against poor performance or business failure begins with the recognition that the sum and substance that creates good performance is housed in the firm's key people. And it is not just the obvious ones, like those whose projects are always profitable. There were a number of retired in-house construction accountants whose less-skilled replacements caused failures. Replacements for departed marketing people have failed to bring in enough profitable business for their companies to survive. There are numerous failures or serious underperformances following the loss of key middle managers in various departments. Most construction company underperformance or failures can be traced to the succession of someone into a position they are unable to perform well at. Once what might be lost if a person leaves a firm is eliminated, how to transfer or cope with it is usually obvious.

Institutionalized Knowledge

Institutional knowledge includes proprietary and general business knowledge processes, methods, and systems used in the business that are available to all employees that need them and are recorded or formalized in some format. Startups and smaller construction companies generally have little or no institutionalized knowledge because everything about the organization, such as proprietary knowledge, processes, and systems, originated with and is housed with the founder. Smaller firms tend to have few if any senior managers other than the founder that use the knowledge. As organizations expand towards mid-size, they begin to have departments and/or divisions that require additional leadership and senior positions that need the knowledge and information to function. As a business grows, it needs to expand its processes and procedures, and new hires bring increased knowledge, processes, and systems into the organization.

If leadership styles and processes and knowledge remain informal, they remain *housed* in the individual senior managers who developed or learned them. Growth requires formalizing processes and procedures to ensure consistency, and these need to be unsimulated. This becomes a continuing activity of recording changes or enhancements to processes and procedures, which effectively institutionalizes all new corporate knowledge. Even with institutionalized knowledge, there are occasionally exceptional employees who exceed all expectations and *create* new or modified processes or procedures. Frequently these are thought to be improved productivity when they are actually *new knowledge*. Often the employee does not recognize that their method is actually new knowledge. Any productivity improvement may be an improvement in knowledge which, of course, needs to be recorded.

Institutionalized knowledge is different in each firm and not necessarily captured appropriately by processes or procedures. The knowledge base is simply the information that a firm has at any given point in time that everyone is using. This ensures consistency, which improves efficiency. Too often contractors resist change by defending the status quo with *that's the way we do it.* Firms that lack institutional knowledge usually undervalue consistency and have managers doing whatever seems best for each. This works fine until it doesn't. Consistency is a hallmark of successful firms and critical to the management of risk. Those that say that you must take risks to get ahead are never fans of institutionalizing knowledge or consistency. With institutionalized knowledge and consistency, employees are generally encouraged to strive for improvements.

Consistency should never prevent continuous improvement. Some mature construction firms have made the mistake of prioritizing sticking with "what works for them" over continuous improvement. Consistency has a limited duration because change is required to keep up with improvements in tools, equipment, and materials. To remain competitive, being willing to change is not enough. We need to be actively looking for the next innovation. To highlight how critical staying current can be, this comes to mind: Quite some time ago, as a very young contractor, I was sitting next to an old, very large, successful contractor at a presentation about desktop computers when they first came out. I said to this contractor, whom I admired and respected, *What's it going to cost us to learn to use these things*? I was relieved when he responded: *Don't worry kid, computers will never come into construction.* As if readers did not have enough to do, add keeping up with technology as a very real key to success.

Transferring Knowledge

Determining *exactly* what a departing (or departed) senior manager does and the knowledge and skill sets they used can and often should alter the person who may already have been selected to replace them. It is common for top management to assume a manager is performing their *job description* when that person is doing a lot more than their one- or two-paragraph job description. Some long-term managers are doing nothing even remotely associated with their title or old job description. This is common and means that top management has limited or no detailed information about the skills and knowledge involved. If that is the case, then a selected successor may or may not be the correct selection or even a qualified candidate. While identifying successors is critical, it is often passed to the employee with seniority. Years of service are not a measure of abilities, qualifications, and skill sets.

The exact delineation of what a person does day-to-day and what skill sets and knowledge they use to accomplish it are the first steps in succession and have nothing to do with their job description. It's not particularly complicated, but it takes time and effort and should be accomplished well before any successor is even considered. The succession process takes years, not months, because the next steps are selecting a successor, developing a knowledge transfer process, and gaining on-the-job experience, which means working alongside the departing managers for months, or longer, depending on how senior the position is. Some will think this is extravagant or exaggerated, but consider the consequences of selecting the wrong person for a senior management position in a small or mid-size construction enterprise. To clarify this point, readers are reminded that a large portion of construction business failures have been a direct result of successions of senior positions.

Succession in small companies is difficult because there are fewer people to choose from, and introducing outsiders into senior positions is difficult. It is a little easier in mid-size companies with more eligible people, but the discontentment of those not chosen can complicate matters. For example, in a very successful mid-size company a younger well-educated manager was selected for president, and several senior executives left the firm within 18 months. The appropriate approach to succession is to announce a succession intention widely within the organization as soon as the decision is made. Important internal activities with a company affect everyone and should not be secret which always cause harmful rumors anyway.

Succession is a *natural* business occurrence and not a negative event until secrecy makes it seem like one. For some reason small and mid-size companies become convinced that strategies, financial information, and the like should be closely guarded secrets, which is not so. All of the distressed construction companies that I helped save eventually exceeded national average profitability for similar firms. All of them share their annual financial statements with all of their employees, which includes public access. Readers who are shocked should ask themselves what their competitors could do with access to their annual financial statements. If you earn higher profit margins than theirs, they might be envious. If you earned less, they would probably be happy. Either way, it does not change their behavior, just their attitude. Harmless from a business perspective, and certainly no reason to give up the huge employee motivation, productivity, and loyalty you derive from it. Do you treat your employees like team members or day workers? They know the difference.

Like all planning, succession decisions are improved when more people are involved in the process. Contractors and senior managers often believe that by virtue of their position and seniority, they are best qualified to make this important decision. They may know more about the company, but they can never know everything—particularly about employees. The higher a manager moves up in an organization, the more they become disconnected from the *employee rumor mill.* The quality of succession decisions is directly proportional to the number of middle and top managers who participate in the decision. It supports teamwork, fortifies succession success, and minimizes or prevents negative reactions associated with successions.

There is no *perfect* successor, but broadening internal participation adds important information to a *pros-and-cons* list and increases the potential for success. The serious risk associated with key employee succession cannot be eliminated. It can only be minimized by recognizing the magnitude of the risk, taking it seriously, planning the process well in advance, and executing the plan, which could take years. There have been successful companies that could not find a qualified successor for the departing contractor that had no choice but to sell or liquidate. More than one hard-working senior leader had to be replaced with two younger generation managers who were qualified but could never be assumed to work the long hours the *old-timer* did. Success in addressing a succession event is improved by moving it from *important* to *critical.*

Non-Transferrable Talents

There are talents or abilities possessed by a manager that are not transferable and as such will be lost to the organization when an individual retires. If any of these are identified, the next step is to attempt to define the impact this loss of *substance* may have on future processes or decisions. The reason I use the words *attempt* and *may* is because some of these talents or abilities are nebulous, vague, or ill-defined. If you can *define* the impact the loss may have on future decisions, then try to determine if there is some way to compensate for the impact. It is difficult to be specific about solutions because these non-transferable abilities, while not necessarily rare, don't ever seem to be the same twice. An example will help clarify.

A highly specialized contractor installed very sophisticated medical equipment. To win a nationwide installation contract, the contractor had to agree to have people available to service the equipment for ten years and renewable for another ten years. The service compensation was very lucrative, but the equipment was so complicated that the contractor had to form a service division and train his people to troubleshoot and repair the equipment. However, problems developed immediately because there were constant complaints about the cooperativeness and behavior of the service people, and legal action was being threatened. My firm was called in. Our investigation indicated that there was never a complaint about the repair work, only complaints about the behavior of the client's service people. We visited many equipment locations and interviewed numerous people who had interacted with the client's servicemen and/or made complaints. (All the service people were men.) More than one customer described the servicemen as "a bull in a China shop," leaving dirt and tools spread around and not leaving the sites as clean as they found them. Other customers complained about loud talking, impolite and uncooperative interactions, and aggressive behavior, and some about foul language.

Several customers said they did not want the same service people to return, and legal action against our client had been threatened a number of times. All of the medical equipment in question was located in hospitals, clinics or large medical offices. There were no separate entrances to the equipment locations, so the tradesmen had to pass through occupied spaces such as waiting rooms, elevators, and long hallways with their tools just to get to the equipment location. They went back through these areas multiple times to get needed supplies, parts, and tools from their vehicles.

Vehicles were often left parked all day in restricted areas at the entrances. These obviously legitimate complaints included all of the servicemen except one, which prompted us to interview client customers in his territory. Everyone interviewed was complimentary about this serviceman, who we also interviewed.

It turns out that of the dozen or so service people in the newly formed service department, the gentleman with no complaints was the only person from the manufacturing division. All of the others came from the construction division, which built the equipment rooms and installed these huge pieces of equipment during the original construction of whatever building they were located in. What became obvious was that these were construction tradespeople that, prior to volunteering to do service work, had never worked around occupied spaces. This is not to suggest that they were undisciplined. Their focus was to get the job done in the simplest and most direct manner possible (*parking closest to the work*). On construction sites they were accustomed to concentrating on the work and *defending their space by* trying to keep other trades (and anyone else) from hindering their progress (*out-of-the-way pedestrians*). Most of those interviewed wish they were back in the field. Old habits are hard to break.

The serviceman from the manufacturing division with no complaints had worked in a hermetically sealed clean room environment wearing a spotless white uniform. The effort required considerable education and training, and he never had to defend his workspace or find it necessary to raise his voice. He had been one of those who trained the others on the detailed diagnostic and troubleshooting aspects of the service job. His background was simply different from those who worked in the field. The solution was obvious. Let the servicemen go back to the field and recruit their replacements from the manufacturing division. The talents required for the new service position were not just the ability to repair the equipment. The repair people had to be comfortable in the environment they would be working in. Half of the new service people were women, which should have surprised no one.

Our client expected us to find a way to explain to their customers that they needed to stay out of the way of the services people if they wanted their equipment to work. They did not expect the proposed solution. Often when a key manager departs, they expose the company to a similar problem. Is this person doing anything different than their job description that we don't realize will be lost to us? One firm replaced a VP who managed all the work for a customer that represented one-third of their sales. Without any consideration about what the customer may want, they assigned their most profitable VP to the account. Within a year the customer was lost. No effort was made to match the replacement with the customer. When this was explained to the contractor, he said that it was none of the customer's business who they assigned. Some of these concepts may not be acceptable to some contractors.

Prepare Successors

Succession is a huge challenge, and the cost of it not working out can be disastrous. Even with a diligently selected successor who has worked alongside the departing manager for a couple of years, there is unfortunately no guarantee of success. It is critical to their success that the candidate undertake continuing education in the skill sets required in the new position to strengthen their talents. The first step in the design of what I refer to as a multi-year *Professional Development Plan* is an evaluation of the candidate's current knowledge and understanding. Unless there is someone in the organization with some educational background or specific experience, this may require some outside assistance. A question about the candidate's current knowledge and experience is helpful. Below is a simple example which readers can add to and change to fit their needs.

Skills Inventory (Confidential)

Please return the completed form

Date: ______________ Name: ______________ Job Title: ______________
Number of Years in Construction: ________ Years in supervision: ________ Years with Co.:________
Highest grade of school completed: ________ Degrees awarded, if any: ________

Other studies-job and non-job related: courses, seminars, conferences, self-study, etc.

Please list, use additional space as needed

Professional or trade qualifications (PE, Journeyman, operator, certified welder, safety certifications, etc.)

Please list, use additional space as needed

Please add any additional comments concerning your education and training that you feel should be included

Use additional space as needed

Please rate what you consider your relative strengths in the following areas:

Technical Aspects of Construction

	Weak				Average			Strong		
Technical field work	1	2	3	4	5	6	7	8	9	10
Field supervision	1	2	3	4	5	6	7	8	9	10
Productivity management	1	2	3	4	5	6	7	8	9	10
Short range planning	1	2	3	4	5	6	7	8	9	10
Understanding scope of work	1	2	3	4	5	6	7	8	9	10
Project management	1	2	3	4	5	6	7	8	9	10
Budgets and cost control	1	2	3	4	5	6	7	8	9	10
Managing change orders	1	2	3	4	5	6	7	8	9	10
OSHA regulations	1	2	3	4	5	6	7	8	9	10
Accident investigation	1	2	3	4	5	6	7	8	9	10

Please rate what you consider your relative strengths in the following areas:

Estimating and Bidding

	Weak				Average			Strong		
Estimating our work	1	2	3	4	5	6	7	8	9	10
Estimating other-trades	1	2	3	4	5	6	7	8	9	10
Getting sub bids	1	2	3	4	5	6	7	8	9	10
Pricing work	1	2	3	4	5	6	7	8	9	10
Putting profit on the work	1	2	3	4	5	6	7	8	9	10
The bidding process	1	2	3	4	5	6	7	8	9	10

Organization and Operation Procedures

	Weak				Average			Strong		
Scheduling techniques	1	2	3	4	5	6	7	8	9	10
Subcontract administration	1	2	3	4	5	6	7	8	9	10

Subcontractor management	1	2	3	4	5	6	7	8	9	10
Purchasing process	1	2	3	4	5	6	7	8	9	10
Cost reporting	1	2	3	4	5	6	7	8	9	10
Equipment selection	1	2	3	4	5	6	7	8	9	10
Equipment utilization	1	2	3	4	5	6	7	8	9	10
Supervising others	1	2	3	4	5	6	7	8	9	10
Business development	1	2	3	4	5	6	7	8	9	10
Motivating others	1	2	3	4	5	6	7	8	9	10
Teaching others	1	2	3	4	5	6	7	8	9	10
Understanding discrimination	1	2	3	4	5	6	7	8	9	10
Disciplining others	1	2	3	4	5	6	7	8	9	10
Coaching others	1	2	3	4	5	6	7	8	9	10
Understanding group dynamics	1	2	3	4	5	6	7	8	9	10
Understanding financial data	1	2	3	4	5	6	7	8	9	10
Fundamentals of business	1	2	3	4	5	6	7	8	9	10

Please rate what you consider your relative strengths in the following areas:

Business and Personal Skills

	Weak			Average			Strong			
Personal efficiency	1	2	3	4	5	6	7	8	9	10
Being organized	1	2	3	4	5	6	7	8	9	10
Management skills	1	2	3	4	5	6	7	8	9	10
Leadership skills	1	2	3	4	5	6	7	8	9	10
Accounting skills	1	2	3	4	5	6	7	8	9	10
Computer proficiency	1	2	3	4	5	6	7	8	9	10
Activity planning	1	2	3	4	5	6	7	8	9	10
Business planning	1	2	3	4	5	6	7	8	9	10
Strategic planning	1	2	3	4	5	6	7	8	9	10
Customer satisfaction	1	2	3	4	5	6	7	8	9	10
Communication skills	1	2	3	4	5	6	7	8	9	10
Understanding body language	1	2	3	4	5	6	7	8	9	10
Presentations to peers	1	2	3	4	5	6	7	8	9	10
Public speaking	1	2	3	4	5	6	7	8	9	10
Writing skills	1	2	3	4	5	6	7	8	9	10
Self-discipline	1	2	3	4	5	6	7	8	9	10
Working alone	1	2	3	4	5	6	7	8	9	10
Working as a team member	1	2	3	4	5	6	7	8	9	10
Negotiating skills	1	2	3	4	5	6	7	8	9	10
Personal goal setting	1	2	3	4	5	6	7	8	9	10
Problem-solving skills	1	2	3	4	5	6	7	8	9	10
Dealing with negative people	1	2	3	4	5	6	7	8	9	10
Personal time management	1	2	3	4	5	6	7	8	9	10
Balancing work and family	1	2	3	4	5	6	7	8	9	10
Control of your career	1	2	3	4	5	6	7	8	9	10

Please list areas you would like to study or improve in, both professional and personal

Return to:		

Discussing the results of this questionnaire with a candidate is helpful in getting to know a candidate and to gain insights into their attitude, intentions, and motivations. This information is used in an evaluation of a candidate's strengths and areas where improvement might be appropriate. The next step is the development of a Professional Development Plan for their advancement into the new position and responsibilities. Succeeding in these programs is critical and failure can be enormously costly. The effort is not just advisable it is not even optional because the alternative is never worth the risk.

Sample Five-Year Professional Development Plan

Name: ____________________

Date: ____________________

Introduction

A multi-year professional development process, by its nature, will also impact personal development. To comprehend the power of three- to five-year-accelerated Professional Development Plans one should understand that in three to five years a person could completely re-make themselves if they cared to. While this is not the primary objective of this plan, most people want to advance themselves over time, and with this process you have the opportunity to design the professional construction person you want to be in the future.

A formal Professional Development Plan is intended to advance a person well beyond their current level of understanding through accelerated learning over and above the work and life experiences that everyone gains each day. It is up to the individual to decide which areas they wish to advance—both professionally and personally. You determine the skills and knowledge you wish to enhance or decide which new skills or knowledge you want to possess.

Skills Inventory

To establish a base line, current skills and interest need to be understood. The skills inventory you completed were used in evaluating current skills, aptitudes and interests and provided guidance in the plan development. We will meet to go over the skills inventory and to discuss which areas you choose to advance in and how you can get there in the shortest amount of time. A schedule of learning activities will be developed to accomplish your goals after which you will need to decide how much time and effort you intend to devote to it. The plan will rely heavily on reading and independent study along with selected seminars and possibly short courses. You should set your goals in harmony with your aspirations and the time commitment you wish to invest in your future.

You will be investing considerable time and energy to accelerate your growth, so your thoughts and feelings are critical to the initial decisions and should influence the content and "texture" of the plan. You and I will discuss alterations to the plan to accommodate your requirements now and as you progress. To be effective this plan needs to be specific to your needs so your input at this stage and over the life of the effort will have significant influence on where we go from here.

Reading

Life-long learning is a commitment to continuous advancement in one's professional and personal life and is directly proportional to a person's determination. The world body of knowledge is captured in books and the ultimate recommendation is to create an attitude and/or change behavior so that reading becomes a permanent habit. Many people avoid reading non-fiction text material because they consider it boring or hard to get through. However, when something you learned from a book solves or avoids a problem, your job becomes easier and when the reading effort pays off it seems a lot less difficult. Serious managers read as many as six or eight non-fiction books a year. I recommend a minimum of three such books a year after this plan is completed as part of continuing life-long professional growth.

To kick start the learning process and to maximize achievement from initial efforts I suggest we start with some easy-to-read booklets that correspond to the subjects we discussed. I suggest that you or the company buy these books because you will want to mark them up and will need them for a future reference library which is addressed in detail below.

PERSONAL DEVELOPMENT:

- ***Body Language; Juluis Fast***—This is old and basic, but I like it. If you find a later text on the subject It will probably be as good.
- ***Unlimited Power; Anthony Robbins***—Good for self-motivation

LEADERSHIP:

- ***Leadership Secrets of Attila The Hun***; Wess Roberts—a classic
- ***Seven Habits of Highly Effective People; Steven Covey***—a classic

MANAGEMENT:

- ***The One-minute Manager; Kenneth Blanchard***—a classic easy fast read.

BUSINESS DEVELOPMENT:

- ***Raving Fans; Kenneth Blanchard***—another easy to read classic.

Personal Resource Material

I strongly recommend that you begin a personal reference library and get into the habit of reading with a highlighter in hand and highlight phrases or sentences that interest you or facts you may want to remember or refer to later. Make notes directly on the pages. In the event you come across words you are unfamiliar with, this can also be a vocabulary-building exercise. Most think we will remember what we read if it is informative or interesting, but it escapes us over time. Highlighted and marked-up books you have read are a tremendous resource in problem solving. You can refer to them throughout your life to recapture information you can't quite remember but know you have seen somewhere.

Trade Publications

Another valuable source of information is trade magazines and journals. I suggest ENR at a minimum and at least one business magazine such as Forbes, Business Week, or any of the many others. They do not have to be read carefully but should be scanned for articles and subjects that you find of interest. Even then the articles need not necessarily have to be carefully read to be of value. If you have an interest in the subject but not time to read it, tear the article out, and file it under general headings like equipment, tools, products, management, leadership, etc. Use a system of file headings that makes sense to you, and this becomes a valuable personal reference source. Then when working on a subject or problem, you can recover information you already have, providing data and insights into the topic at hand.

These are good habits to get into even if you don't see the value up front. They are proven techniques that work similarly to keeping the books that you have marked up. When you start the last book, advice and more recommendations will be sent.

Self-Study

Many skills, particularly communication and negotiating skills, can be advanced in self-study. After you read the body language book, put into practice what you learn. Become more conscious of studying people more thoroughly through their actions as well as their words when you interact. This is an excellent skill to perfect. It is valuable in managing, marketing, negotiating, and presentations. After you try this for a while, if the subject is more than a passing interest to you, another book on the subject will be included in the next selections. After that we will move on to the study of group dynamics.

We may want to delay for a year or so, but you should consider becoming active in a trade association that your company is a member of. The reason is that associating with other construction professionals is a learning experience and a part of the area construction community.

Seminars and Conferences

These need to be selected as they become available. You attend two or more times a year. Most are announced three to six months in advance, so it is an ongoing selection process. You should be looking for programs of interest to you.

Short Courses

There are a number of one- and two-week courses available that may form a part of this plan, but not in the first year and not until definitive progress is made. These are appropriate but time-consuming. They are important for this effort but not initially.

Long Courses

For very senior positions I have recommended long advanced programs and in-residence study such as the Harvard Business School Executive Program and the Penn State Executive Program. This surprises most people, but the results have been exceptional.

Follow-Up

It would be helpful to prepare a written or oral report to management on your progress about three times a year. It is appropriate to do a self-assessment and discussions with management as an effective assessment of progress and future steps.

A Continuing Process

A multi-year Professional Development Plan is not set out in its entirety because the pace of professional and personal development is set by the candidate. It is also dependent on the availability of learning opportunities. The plan will be updated periodically (at least every six months), in consultation with the candidate. Management is available for assistance and consultation, but most of the interaction should be prompted by the candidate as they see the need.

Succession Plan Management

After several months into a succession plan, it is appropriate to measure progress. The following short questionnaire is a method of self-evaluation and includes a memo to managers who will be evaluated.

To the succession candidate: The company is investing in your future, and your thoughts on your progress form the basis for evaluation and improvements to your professional development.

Skills Inventory Progress		
Please return the completed form		
Date: ____________ Name: ____________		
Please answer the following questions on a scale of 1 to 10, with 1 meaning little progress and 10 being excellent progress. Add any comments that you feel apply.		
	Score	Comments
Your progress in being introduced to client contacts		
Your progress in establishing relationships with new clients		
Your progress in managing and improving existing client relationships		
Your progress in establishing and improving relationships with designers		
Your progress in establishing and improving relationships with vendors		
Your progress in establishing and improving relationships with union contacts		
Your progress in understanding how to assign superintendents		
Your progress in understanding mobilizing and moving equipment effectively		
Your progress in being instructed in production and process knowledge		
Your progress in being instructed in project risks		
Your progress in pre-bid evaluation, pricing and risk assessment		
Your progress in managing moral and effective team building		
Your progress in becoming a safety champion		
Your understanding of areas that you need to improve in		
Your progress in knowledge growth through your PDP learning activities		
Your progress in mastering negotiation skills		

Succession to President

The responsibilities of the president of any business are huge, particularly in the construction business. Where second-in-command executives have major but specific and specialized issues to deal with, the range of entrepreneurial responsibilities of total or partial business owners is massive. In very large firms, the president can sometimes be a figurehead, but in small and mid-size companies they are responsible for everything. Producing the work for a profit is difficult, but not nearly enough to remain in business. Before we review the business issues a construction company president is responsible for, we should clarify that many presidents are not aware of these responsibilities. Some say others are responsible, which does not remove ultimate responsibility from the president. Others say they are not qualified in certain areas, which of course does not alter their responsibility.

The following are business functions that the president is responsible to ensure are executed well. If they just had to ensure these were executed, they could be delegated, but executed well demands an understanding and evaluation, which means supervision. Arguments about this are almost exclusively semantics. Some may say one or more is not important, but the measure of importance is that all of these have contributed to construction business failures. Therefore, these are each critical construction business functions.

- Accounting: *use of financial information/planning, cash flow management*
- Banking and Borrowing: *use and management of relationships*
- Business Opportunities: *evaluation, strategies to enter or get out of*
- Change Orders: *management, pricing*
- Computer: *technology, appraisal, selection, acquisition, usages*
- Contracts: *negotiations, administration, understanding of contract law*
- Crisis: *management, avoidance,*
- Customer Service: *client maintenance, customer relations, networking*
- Designer/Inspector: *managing, including appropriate relationships*
- Disputes/Claims: *avoidance, negotiation, resolution*
- Equipment: *evaluation, purchase, tool selection, maintenance, management*
- Estimating: *strategies including adjusting pricing to circumstances*
- Labor: *management, negotiation, understanding prevailing regulations*
- Leadership: *including employee selection and management*
- Marketing: *advertising, networking, and screening potential customers*
- Marketplace: *understanding, evaluation, market evolutions, and cycles*
- Motivation: *team building, earning loyalty, including compensation and benefits/perks*
- Organization Structuring: *maintenance and updating*
- Overhead: *management and control*
- Planning: *business, short-term tactics, and long-term strategies*
- Productivity: *management, operating policies, production standards, costs*
- Professional Development: *of self and employees*
- Project Selection: *screening, market and competition analysis*
- Risk: *recognition and management including insurance evaluation/selection*
- Safety: *management, administration, OSHA, all prevailing safety regulations*
- Subcontractor/Vendor: *screening, selection, management*
- Surety: *use and management of relationships*

A number of contractors have claimed that they have people to do some or most of these things. In effect, all the actual effort to accomplish just about everything on this list is performed by others. However, that does not alter the fact that the contractor/owner/president must know what must be done, assign the work, and be qualified to evaluate the results. The contractors who say they don't know anything about one or more of the items on this list are exactly my point. *Business competence in the construction industry is weak.* If both this statement and the failure rate are true, then we have uncovered the problem. All we need to do is recognize the solution, which is *informed* and preferably *strong* leadership. The starting point for that is to understand the personal *attributes* that contribute to business success:

- Ability to listen attentively
- Generosity of spirit
- Realistic self-confidence

- Curiosity
- Honesty
- Candor
- Assertive, not aggressive
- Sense of humor
- Clarity
- Empathy

Few individuals possess all these attributes, and those who have some of them have them to varying degrees. These are the attributes that research by psychologists and management consultants found prevalent in successful leaders. I have encountered the argument that these won't work in construction, with some claiming that "toughness" is missing. These types of comments were mostly from mature contractors. Toughness was widespread in our industry some decades ago, and there should be no need to convince readers that today's workforce does not respond well to toughness. In fact, toughness may be partially responsible for fewer young people selecting the construction trades as a career choice.

Family Members' Succession

This subject was saved until last because it is hugely controversial and can be painful for some. The ability to successfully manage a construction business is not hereditary (Figure 12.2). In the past, many in our industry disagreed with this, and some still do. Family succession is, of course,

Figure 12.2 Family succession is common in construction, but success depends on qualifications, not heredity.

optional, but believing that it is the natural and therefore always correct thing to do is not up for debate. The astronomical failure rate of startup construction companies clearly demonstrates that the average person won't succeed. The long list above of functions a contractor must manage eliminates most people. Few have the intensity, even audacity, to try. Put another way, most are far-sighted enough to avoid the enormous stress, long hours, titanic risk, and inadequate return on effort or investment. In brief, running a construction firm of any size is arduous and demanding, and few would describe it as pleasant or gratifying.

There are very few people qualified by education, experience, and temperament that have the potential for success in running a construction company. The *thankless* position described above should not be forced on anyone. Contractors, of course, say: *They want to run this company*! Or: *They worked summers in the field through high school and college*! Or: *I taught them everything I know*! Or: *They were brought up with construction around the dinner table*! The only thing being brought up with construction around the dinner table teaches you is bad table manners. Working every summer from eighth grade through college as a laborer or in the trailer is not much of a head start to president. If it is not *in the blood*, our offspring have the same odds of being a contractor as any apprentice. It is a disservice to them and to the industry to make them think otherwise.

Summary

There are almost 1.5 million construction companies in the United States. That is 1,500 times the number of publicly traded companies in the country. The construction industry has a huge number of leadership successions every year. If leaders of construction companies serve for 20 years, which is considerably long for firms other than startups, that would be 75,000 successions a year. That does not include the senior executives that depart and must be replaced, which would be orders of magnitude more succession. This departure of talent and knowledge that must be replaced is massive exposure for the industry. It may seem like it should be of less concern for individual companies because it happens to them less frequently, but that is not the case. The eventual loss of senior talent is a *certainty* for all companies. The unexpected untimely loss of senior talent is a *potential* for all companies. Therefore, succession planning, which may have seemed optional before reading this chapter, is in fact *required* by risk management principles.

13 Joint Ventures, Always Risky

Introduction

Contractors seem to be driven by growth to one degree or another and whether intentionally or unconsciously, the majority claim otherwise. None seem to accept the research that confirms the direct relationship between growth and business distress and failure. The only safe haven for growth in sales, as mentioned in previous chapters, is 15% or less annually. Construction companies growing faster than 15% show a statistically significant reduction in profitability with a small increase in business failures, At any percentage over 15% and up to 50%, the amount of reduction in profit and the small increase in failures increase almost proportionally. At rates of growth over 50%, the percentage of profit continues to reduce while the rate of failure increases. Over 50% rate of growth, the failure rate continues to increase and should be avoided altogether.

Readers may ask, if this chapter is about joint ventures why begin with a reminder about rates of growth? Because the primary motivation for joint venture almost always revolves around growth, and some have even suggested that joint ventures are a form of *safe* growth. This could not be further from the truth. As a reminder: Growth in the size of projects is the first and biggest of the five common causes of construction business failure. Change in location of projects is the second common cause, and change in type of project is the third. These three common causes of construction business failure also happen to be among the primary motivators for forming construction joint ventures. Add to this that the formation of joint ventures normally includes key personnel from two different *partner* companies and a lot of new employees, which is the same as the fourth element of failure: *changes in key personnel*. It would be easy to *weave in* the fifth common cause, a lack of managerial maturity, but I will resist the urge.

Joint Venture Defined

There are numerous descriptions of *Joint Ventures*, but few useful or informative *definitions*. Interestingly, in Black's Law Dictionary (4th Edition), the listing for *Joint Venture* says: *See Adventure* which tells you nothing. A common description of joint venture is:

A collaborative business arrangement where two (or more) businesses combine resources, expertise, and knowledge to achieve a common goal, usually a specific project.

It is assumed by most that the partners in a joint venture project share a proportional share of the costs, profits, and liabilities and many joint venture agreements actually say that. However, in all cases I have been involved in if one joint venture partner failed, the courts have said that the remaining partner (or partners) are *jointly and severally* liable for 100% of any liabilities

DOI: 10.1201/9781003500315-14

(loss) incurred. The surviving partner is responsible for any loss by the joint venture no matter which partner may have caused the loss. It makes no difference if the surviving partner did not know they would be liable or disagree that they are liable. Readers who have trouble believing this should check with their attorney. We know that most contractors don't read every contract/agreement they sign. Many claim they would not necessarily understand the legalities anyway. Most have their attorneys read them, but that seldom means that the details of every page were then discussed.

Profitable or Not

All of the motivations and reasons for forming a joint venture discussed here were reported by joint venture contractor partners, both contractors with successful joint venture experience and others who experienced unsuccessful joint ventures. Some of the motivations were reported by contractors with a series of successful joint ventures before a disastrous one. Those with negative experiences were unanimous that they would never enter another joint venture. Unfortunately, some of those businesses did not survive their negative joint venture experience. Conversely, all contractors with positive experiences praised joint ventures and were committed to taking advantage of any joint venture opportunities that came their way.

Without more data one could conclude that joint ventures can be either positive or negative. This raises the question: is there a screening method that can be used to select only successful joint ventures? Because some contractors had multiple successful joint venture before a losing one suggests there is no screening method. That this group quit joint ventures after a single loss suggests that they each must have blamed their loss on the *joint venture structure itself*. If they blamed themselves or their partner, there would be no reason to quit joint ventures.

A study of hundreds of successful and unsuccessful joint ventures revealed there are no clear differentiators between the two. No patterns or behaviors that were obvious causes of success or failure. Some insight was gleaned by continuing to collect data for more than 25 years after the initial study. Numerous large contractors, with a number of successful joint ventures, also quit doing joint ventures. The reasons given were that joint ventures were just too risky and/or their bonding company would not continue to support joint ventures. Most said they were no longer comfortable putting their future in a venture they could not totally control. More than one said they were no longer interested in huge projects. It is difficult to draw any definitive conclusions from this sampling, because there is not a statistically significant amount of data, but intuitively this information is not very supportive of joint ventures (Figure 13.1).

Types of Joint Ventures

Equal Partners

The motivations for forming joint ventures may be helpful in evaluating their value. The classic reason is that a project is available in a contractor's area of influence that is too big for the contractor's organization. For example: A contractor whose biggest prior job and single project bond limit is $10 million dollars has an interest in a $20 million dollar project. The contractor decides to try to form a joint venture with a contractor of similar size. The problem here is that a construction firm attempting a project twice the size of anything they have built is by any definition, *not qualified*. Some will argue that perhaps they can build it. *Qualified* in this context means *having successfully built it before*, which means similar in type and size. Some then argue how do you grow, and the answer is slowly by small increments. This is not worth arguing about because it is an industry standard. Therefore, by definition, this joint venture will be two contractors, neither

Figure 13.1 Joint ventures often appear attractive as a path to growth, but they rarely reduce risk and can amplify it.

of which is qualified for this size project. In addition, neither can qualify for a $20 million dollar bond, but with cross guarantees one of their bond companies may issue it. Some have suggested two $10 million bonds are effectively the same as a single $20 million bond, but few project owners are informed enough to accept two separate bonds nor will a knowledgeable bonding company issue them.

Large and Small Partner

A large contractor with a big project in an area they don't have a presence in often looks for a local contractor to joint venture with. There are a number of reasons such as assisting in performance of the work and for a connection to local subs and supplies. Bonding capacity is not usually an issue. The large contractor often supplies field management, and the smaller contractor becomes responsible for getting the work done. These are not necessarily 50/50 ventures, and the smaller partner may seem more like a subcontractor. A common difficulty with this arrangement is the dominance of the large partner, and the smaller partner is often over their head attempting a project size way outside their experience. The small partner may not be accustomed to the slow pay processes typical of big projects. This extends to the subcontractors and suppliers the smaller partner attracted to the job. With limited negotiating leverage, the small partner's profit share may be an issue, and it is not unusual for the smaller partner to realize they have little or no bargaining power at all. It is difficult for the smaller local contractor to anticipate it, but their business is typically disrupted completely because they must focus on the project that is too big for them to begin with. Unequal size and authority in this type of joint venture means the smaller partner has too big of a project with a difficult boss.

Type of Work Ventures

There are joint ventures based on types of work that have at least the possibility of working out. Take for example a baseball park, track-field competition area, or similar projects. These have buildings, bleachers, etc., and a huge amount of earthwork, landscaping and parking lots. The two distinct types of work would require a building contractor to get sub bids for the groundwork, which they typically have limited experience with, but is half the project. Or, a groundwork contractor to get sub-bids for the building half of the project, which they have limited experience with. A joint venture with the two contractors would likely result in better pricing and greater control of the work if they get it. It helps if the two contractors are of similar size and sophistication and if there is some level of trust, because trust is always a concern. Bonding is always a problem, and the fact that neither can perform the work of the other in a crisis is a negative to this approach.

Joint Ventures of Convenience

It is not uncommon for large contractors to have a steady client ask them to build something of a type of construction they are not experienced in or in an unfamiliar area. If the project is in an area where there are no contractors big enough or experienced in the type of work, local assistance is a problem. A small less qualified contractor will be needed to attract the necessary local labor and subs. Often none of the small local contractors are even qualified to deal with negotiations let alone build the project. While a subcontract arrangement may seem logical, a joint venture partner always has a deeper commitment—even junior partner. Unequal joint ventures are problematic in a number of ways. In this case the contractor with the project is the majority partner. The small contractor that will be performing all the work has no idea what to ask for. This is an opportunity to build a project, which is a strong motivator to accept whatever percentage is offered. The problems are the smaller partner is over his head and the larger partner's field management team are not prepared for the quality control issues that are in their future. Joint ventures of this nature have limited odds of success.

Bonding Capacity Ventures

There have been joint ventures formed for the sole purpose of combining two contractors' bonding capacity in order to bond a project that neither can bond themselves. Oddly enough there have been a number of these arrangements inspired by bonding agents. A contractor's bonding capacity is said to reflect what a qualified bond underwriter calculates is the dollar value size of project the contractor can produce and finance. A surety underwriter that approves such a joint venture seems to be suggesting that the two contractors combined financial capacity satisfies the surety's financial criteria. Neither contractor has built anything this size, so the fact that the bond is approved must mean that all concerned in issuing the bond believe that the combined production capacity is automatically doubled. Or they think that their building experience somehow increased with the increased financial capacity. Or that building capacity experience is less important than financial capacity. None of this alters the fact that a joint venture of two contractors not qualified to build a project results in a joint venture of contractors building a project that neither contractor is qualified to build.

Joint Venture vs Subcontract

An alternative to joint venture is for the contractor with the greater portion of the work to subcontract the portion of the work that they considered a joint venture partner for. This can be accomplished with an *open book* process, which builds trust. It is simply less complicated. For example, there is a road contractor that does not build bridges and bridge contractor that does not build roads who regularly work together. They are often interested in projects that have both bridges and roads and have cooperative arrangements with each other. They each price their portion of the project independently, and whoever has the larger value of work bids on the project, including the other's subcontract price as a subcontract. If they are successful, the smaller portion of the work is constructed under a subcontract. They had been engaged in this cooperative arrangement for many years without dispute. When asked why they don't joint venture these projects, they said they had heard my position on joint ventures and agree that they have too many complications. Their prior joint ventures were clumsy with unclear risks and too much paperwork. Note: Forming joint ventures because of differences in type of work may seem like a good approach, but subcontracting is the more risk-free alternative.

Providing Qualifications to Others

There have been joint ventures formed as a vehicle to provide bonding or pre-qualification credentials to another contractor for a fee or profit percentage. There was a case in an earlier chapter where a large contractor provided bonding capacity to a much smaller out-of-state contractor who was unable to secure prequalification from the project owners and was not bondable. The projects in question were selected, estimated, and performed by the small contractor. They were contracted with and bonded by a joint venture dominated by the larger contractor, who had no connection with the work. In most cases the large contractor never saw the estimates or the projects themself. The author became familiar with these cases after the smaller contractor became insolvent and bankrupt. The larger contractor was jointly and severally liable for the loss regardless of what their joint venture agreement said. There were quite a number of similar cases like this. Some of the large contractors were able to pay for the subsequent loss, and some were forced into bankruptcy.

Joint Venture Recommendation

Successful and unsuccessful joint ventures, just like jumbo-sized construction projects, all appear successful from the outside; and also to everyone who worked directly on them while construction was in progress. If the *appearance* is incorrect, it is usually known to only a very few senior partners. Celebrations and ceremonies at project completion always seem to announce that the success of the project was the combined effort of all parties involved. Any follow-on unsettled contractor claims are considered *misunderstandings*, the results of which will be determined many years later and announced in fine print. The point here is that financially successful and failed joint ventures all look alike. There are no obvious tell-tale signs of *unsuccessful* joint ventures. The reality is that very few even highly placed senior executives of either party to the venture actually know if a joint venture project was a financial success or not. And if a joint venture was unsuccessful, the partners are not going to tell anyone but perhaps their attorneys. It would be bad for their and the firm's reputation. Incidentally, the same is usually true for all jumbo projects, including those that are not a joint venture.

Because there are very few admitted failures of projects having been completed via joint venture, they have a positive status and appeal within the industry. This status and appeal is not data driven. It is the *opinion* of industry observers and the professionals involved in the venture. The majority of industry professionals have never been the senior executives of a joint venture partner. There is no statistical evidence available as to the success of joint ventures, so we are left with the limited research findings combined with years of first-hand experience into the subject of the primary causes of construction business failure. Many, perhaps the majority of joint ventures are formed by firms that are too small to build and/or finance the project alone. Once the venture is formed, the firms are still too small to build or finance the project. There is no mechanism whereby a joint venture could reduce this risk. It is exactly the same risk factor for type of work. A firm that states they can't do a type of work and then enters a joint venture that includes that work is signing a contract to perform that work. A joint venture agreement cannot and does not change, enhance, improve, or alter the firms' qualifications to perform the work in any way.

Is it possible to answer the question, is a joint venture a viable, appropriate, and profitable construction contracting method? The first concern is that all the joint ventures studied violated one or more of the well-known and accepted *five common elements of construction business failure*. The majority of joint ventures are performed by at least one partner for which the project is:

- The largest project ever attempted
- In an unfamiliar geographic area
- A different type of construction that ever attempted
- Is to be built with a majority of new employee teams
- Will by directed by manager who have not worked together

If none of these were true, why then would there be a need to form a joint venture?

If the need for the joint venture is financial, such as insufficient bonding capacity by either partner, then if the one partner fails, both fail. Perhaps a long shot, but if you would not risk your firm in other ways, why would you risk it in a joint venture? If both partners could bond the project alone why joint venture? The reason sometimes given is a desire to share the risk because neither wants to take that much risk alone. This reason makes no sense because from the evidence so far presented, it is obvious that both partners are liable for 100% of the risk in a joint venture. Join ventures don't and can't reduce risk in any way. Because of so many other factors, joint ventures always increase the risk.

The Answer

A prudent businessperson never puts their company at risk and particularly unnecessary risk. Performing the size and type of work in the same location you normally work in with your trusted long-term employees in a joint venture increases your financial risk dramatically compared to performing that same work yourself. I don't know how to make it any clearer.

My recommendation to contractors of any size, anywhere, is never, under any circumstances enter into a joint venture agreement.

Business Risks

There are too many misconceptions about business risks.

- There are business risks—True
- There are unavoidable risks—True
- You have to take risks is business—False
- You have to take risks to make money—False

There are no avoidable risks because if it is *avoidable*, it is not a risk. No businessperson would take an avoidable risk because if the action is avoidable, why would they take it? Too many people think there are logical reasons to take an avoidable risk. Presumably they are thinking there may be some gain by voluntarily accepting an avoidable risk. That logic does not and cannot exist in business. Because: A common definition of *business* is: *An organized effort to produce or sell goods or services for a profit.* A commonly accepted purpose of business is: *The primary goal of business is to generate revenue and profit through the production and sale of goods or services*. These descriptions of *risk* are clear, declarative, and proactive statements that imply nothing about risk. If any businessperson believes that producing any product or work may not result in profit, then by *definition* that businessperson would not produce that product or work. If a person determines to produce that work in an effort to determine *if work generates a profit*, then they are simply *gambling*. Gambling is not business and the work should not be produced. *Gambling* is not *business*.

A businessperson is usually fairly certain that their production will generate a profit. Does that *guarantee* it will profit? No, but at least they feel certain. There is always a possibility that production may not profit, which means it was an unavoidable risk. Only if they were uncertain to start with would it be an avoidable risk. No one should ever willingly take an avoidable risk. That's not business; that's gambling. The argument against this from students is that *you can never be certain*. The response is, then you should never go into your own business. When a doctor hangs out their shingle they announce, *I can cure you.* When a contractor goes into business, they announce *I can build for a profit.* Are they both certain? If either of them does not believe their announcement, they should find another position. Are they correct? Let's hope so. Are they confident? If they are well prepared, they are confident. Both know that risks exist (which we have also said here). But neither should ever deliberately take any business or medical risk. Doctors are taught not to take risks, and now construction professionals are taught not to take risks. Many who disagree may do better as gamblers than constructors.

Risk-Multipliers

There are unavoidable risks on both the business and production sides of the construction industry. The risks unique to joint ventures are *risk-multipliers* to an already high-risk industry. Both producing the work and accounting for the work are complicated and costly, and mistakes in these areas can and have caused business failures. Similarly, changes in key personnel can cause costly and critical mistakes. All these mistake potentials are built into joint ventures when resources and personnel are brought together from each firm to produce and account for the work. The fact that some of the employees may be long-term, trusted people, when mixed together, are diluted and in effect are a *change in key personnel*. A change in key personnel is of course the second common cause of business failure. Adding new hires into the mix, and it grows into a *massive* change in key personnel. The work will be produced and accounted for by crews and construction professionals with no prior experience working together as a team. They are strangers to each other and a risk multiplier.

A joint venture presents too much *change* to contend with. Because this combined employment is not permanent, special team building or training steps are seldom taken. This is further complicated by the fact that some work groups will be on the project for a limited time, for *their* phase of the work. Joint venture projects have been described as *being built by a group of strangers*. Efficiency and productivity are affected, and seldom for the better. Adding even more *differences*, a huge multi-year non-union project was built by a joint venture of a non-union and a union contractor. Everyone involved seemed to be surprised by the continuous labor unrest. Joint ventures are not *business-as-usual*, but they always seem to be estimated using standard *business-as-usual* pricing. Seldom do joint venture estimates include funding to compensate for the *built-in* and widely known complications and inefficiencies of joint venture production. Pricing for these *exposures* would of course reduce the odds of winning the job. First-time joint venture partners never anticipate the exposures because everyone seems to believe joint venture projects are and will be produced exactly like any other projects. Unawareness and inexperience with joint ventures are risk multipliers.

Another peculiarity of these arrangements is that there never seems to be any small joint venture projects. Joint venture projects tend to be among the biggest jobs either partner has ever attempted. That being the case, it is probably safe to say that most joint venture projects are exposed to the number one cause of construction business failure—change in size of project (by more than 15%). It was necessary to devote an entire chapter to joint ventures because of the huge number of negative exposures that compound each other as *risk multipliers*. To be fair and balanced, all of the positive elements of joint ventures in the construction industry have already been listed in this chapter. If there were any, they would help to offset the negative elements.

Joint ventures tend to be very large projects like energy generation, bridges, tunnels, and highways, and they are sometimes in remote, even unpopulated locations. It is not unusual for very large, remote projects to also be unique or one-of-a-kind projects, which are therefore very challenging to design, estimate, and build. The transportation and storage of equipment to remote areas and the accommodations for manpower at these sites are huge challenges that few contractors have much experience with. Any change in location increases risk, but remote locations are in a class by themselves, with many including a portable clinic, food service, laundry service, temporary utilities, site and personnel security, and the list goes on. Working in a location where a contractor has not worked before (*change in location*) is the second common cause of construction business failure. Determining how long a huge, unique, remote project might take to complete is almost impossible. This means there may be some *guesswork* in the schedule for the project. If the schedule is uncertain, and pricing depends on the schedule, when the *guesswork* finds its way into the pricing, it becomes a risk multiplier (Figure 13.2).

Figure 13.2 Even successful-looking joint ventures can conceal disputes, cost overruns, and financial distress.

Defining Success

Success in the construction business has different meanings for different contractors, which complicates all of the issues covered in this book. For many, successes means growth, or the *If you're not growing, you're going backward* attitude. For others success means building unique and/or highly visible projects and having the company sign all over town. For some it is about winning a race with their competitors. For some it is the thrill of business risks, and for others it's making money. If you own the business, you can define success any way you choose. However, we should all understand that the only viable and realistic definition of business is as stated above: *An organized effort to produce or sell goods or services for a profit.* That's it—profit not growth. If a reader believes that growth is a primary purpose of business, they will not be able to relate to much of this. What is not included in the definition of growth is any mention of *risk*. The risks in business are the same as the risks in life—things outside of your control happen. Hardly anyone deliberately takes life risks except daredevils and race car drivers. Construction people seem to be convinced that you have to take risks to make money. Which is simply not so.

The statement: *If I take this big job, I could make a lot of money* is incomplete without an explanation of the word *could*. The meaning allows for other possibilities such as *could lose money*. The following absolutely true statement separates the businessperson from the gambler. *No businessperson should deliberately make a decision that could cause their company to lose money*. (One contractor responded to this: *You take all the fun out of the business*.) Because it is true that *construction is risky*, people seem to believe that means we have no choice but to take risks. Or worse, that taking risk is just part of the business. That is not what it means. It means that the construction process and the industry have some built-in risks. Built-in risks are *unavoidable* risks.

Too many people have somehow confused the fact that "*they work in a risky business*" with: "*Risk is OK*," or "*Don't worry about risks*," or "*All contractors take risks.*" The construction business has risks does not mean that all contractors take risks. And it certainly does not mean that contractors should take risks. Construction professionals in this industry have got to understand that their job description includes, do not take risks and includes protect the company from any and all risks. Construction professionals have got to stop thinking risk is Ok. Or worse, we are not afraid of risk.

Business Not Experimenting

Another activity that has somehow found its way into our industry is that we should always *be looking for a better way to do things*. Successful construction companies produce the work profitably, safely, and consistently. Managing, supervising, and producing the work does not include *experimenting* with company practices in an attempt to find a better way. Here is an example of construction professionals experimenting with what they thought was a better way. A multi-story structural concrete building under construction had advanced to the placing of concrete on the sixth floor when the entire building collapsed killing several workers below. The investigation included discovering that a supervisor had ordered first-floor temporary shoring to be removed, saying: *We were in a hurry and the concrete above was hard, so I assumed it was safe to proceed.* Process improvement requires a combined effort and analysis by a minimum of production, financial, and safety experts. Process improvement is not a trial-and-error process. Success starts with having standard procedures and enforcing them.

Defining success requires the accurate measurement of performance, which in our industry, particularly in the field, can be problematic. For example: The evaluation of a financially distressed large masonry contractor determined that their method of cost control was the primary cause of their failure. The firm had no standards for how field management measured masonry block and brick production. Field reports of unit costs were the basis for field performance bonuses and also used to estimate and capture future work. It turned out that most of their field production reporting was overstated, and apparently no one in the office ever verified the amounts. On a few projects the total number of masonry units actually incorporated in the work was overstated. On other projects some masonry labor costs were charged to other categories, which of course caused cost overruns in those categories but lower masonry unit costs. And on some projects both errors appeared. The overstated production rates were used to estimate future work, which fueled growth for some time before the firm became insolvent. Business skills are an issue in startups and small firms and to a lesser extent in some mid-size and large firms as well. Some may say no wonder they failed, but for the record, this firm had been in business for decades. Constant growth generates cash flow if not profit. The definition of success is profit, but the measurement of profit depends on accurate field reporting processes.

Managing Growth

Growth is a motivator for joint ventures, and joint ventures are almost always large projects. Decisions to join a venture are usually more opportunistic than part of a long-term plan. Often a project larger than normal for a contractor will not be envisioned until they or someone else considers a possible joint venture. This is not an everyday event because the projects in question are not what this organization normally goes after, and the joint venture process is not the company's normal contracting method. This is true for both first- and few-time ventures. The project is bigger than anything they have built, and they seldom participate in a joint venture. This should be a caution in itself. The contractor then enters into a joint venture agreement with a relative stranger. Together they form a new legal entity that will have a contract with a project owner. The project owner is usually a new client (relative stranger) to one or both of the joint venture partners. (another caution)

A contractor's motivation to joint venture can be tricky. Everyone likes a growth opportunity, so they usually at least explore the process. This is roughly the conversation several venture partners told me they had with themselves:

> What harm can it do. This is not really growing my organization. It is more like an investment opportunity. My joint venture partner and I will meet once a month to oversee this operation, and I'll get half the profits. If we each put a couple of our key people into this thing it will have more field management than I put on my normal jobs so it will run itself. I can spare a day a month and maybe I should open a separate bank account for the profits. With the joint venture employing and paying some of my top key people, I'll be making more money on my regular work. I can't lose. This is a windfall that I need to jump on.

If or when this happens, all other serious considerations or evaluations are put aside, blocked out by opportunity and thoughts about what to spend the windfall on. This results in aggressive pursuit of the project, including sharp pencil estimating, possible concessions in contract language, and the potential for last-minute reductions in pricing. The celebrations after winning joint venture competitions have been known to be excessive. This scenario has been repeated many times. If the joint venture is successful, even if not as successful as anticipated, contractors begin looking for another. That pattern continues indefinitely until a failed joint venture results in bankruptcy or with a contractor that swears off joint ventures for life.

Profit is the only measure of success of a construction enterprise. Any continuous growth at a percentage rate greater than the concurrent rate of growth of the construction market in question will eventually result in a reduction in profit percentage. Many claim they are growing rapidly and maintaining profit margin, but response is to triple-check your work-in-process accounting because it is wrong. Modest growth slightly above the market growth rate can be managed with just a modest reduction in profits. Major growth of 25% or 50% above market growth for any continuing length of time will reduce profits and may just as easily result in loss. The problem is that once the work is under contract, there is no way out, which is why I strongly recommend caution. Before undertaking such growth calculate a worst-case scenario of losing 5%, 10%, and 15%, and if you can't financially afford the worst-case loss and remain solvent don't do it. Informed businesspeople never take risks and certainly not the risk of bankruptcy. It is too much like Russian Roulette.

It is hopefully becoming clear to readers that projects that have multiple proven high risks built into them seem to attract joint or multiple ventures. Some have speculated that these enlightened contractors recognize the risks and attempt to team up with others who have specialized talents and can share the risks. This might very well be true, but does it mean that each partner can accurately *identify* and perhaps more importantly *quantify* these risks. Particularly on a project larger than they have experience with. The size of the project is unique to their executives, managers, and estimators. Sharing the risk is a misnomer with joint and several liability, but the real questions become, can your team *identify all of them* with any certainty and can they put an accurate price on each? It gets worse. If the joint venture partners identify and price all the risks correctly, will they lose the job to the contractors who identified or priced them incorrectly? If success is being the low-bidder, then excellent risk identification and pricing accuracy will not likely achieve the lowest price.

Summary

Missing from this chapter are any positive reasons to form a joint venture to build construction projects. There was none to be found. A serious follow-up effort was made to research the question with knowledgeable construction professionals with experience in joint ventures. Any that had a negative joint venture experience had nothing positive to say about them. Those with positive experiences reported joint ventures were a way to build larger projects than they could alone, earn profit with less risk because of the protection of the venture, and several reported less profit but were satisfied because of the reduced risk. It is a surprise that some very large contractors were misinformed. The risk is actually twice the risk for a half share of the profit. A follow-up study revealed partners that quit after a failed venture included some of the satisfied partners in this paragraph. I repeat my strong recommendation that any and all joint ventures be avoided, period. The better it looks, the further you should stay away.

14 Equipment Costs More than You Think

Introduction

Depending on the type of work performed, contractors range from *equipment intensive* like road builders, and utility contractors, to moderate equipment usage like painters and glaziers. Equipment usage varies by type of work, but the decisions, costs, and management of equipment impacts the efficiency and profitability of all contractors. Construction businesses of every type and size use equipment in the execution of their work except for non-performing general contractors or construction managers. Non-performing general contractors and construction managers perform more like brokers or managers who supervise and direct the work of others. In that role they neither own nor operate construction equipment and are therefore not included in this discussion when referring to *contractors*.

Selecting equipment on both equipment-intensive projects and those with limited equipment requirements results in reduced profits when that equipment turns out to be less productive. The purpose of this chapter is an attempt to convince readers that equipment knowledge, choices, and usage are more critical to success than most construction professionals seem to believe. This is more prevalent with less equipment-intensive trades because of the reduced role equipment plays in their production process. It appears that less dependence on equipment causes less concern about the role equipment efficiency plays in production. In reality the efficiency, and resulting cost of the manpower involved, is dramatically impacted. The point is that this chapter should have the same interest to both those who use a greater amount of equipment as it is to those who do not employ as much equipment. Both equipment basics and usage will be covered.

Ownership Costs

Equipment affects every type of construction company and accounting for its costs is a complex process. Evaluating and recording all costs associated with equipment ownership is complicated by the fact that some costs do not appear on invoices and aren't incurred in the course of operations. These costs, while not easily recognized, will eventually have to be paid. Although various formulas are used to cope with equipment costs, the total costs of ownership are often misunderstood. Efficiency demands that all costs, including those hidden, must be recognized and planned for in advance. There are other equipment concerns, such as whether to buy, lease, or rent; which equipment to invest in; and when to invest.

Equipment-intensive contractors, like road builders, utility contractors, and sheet metal contractors, have to make these decisions often, and they know that their decisions have a long-term and profound effect on their success. Other contractors have the same exposures, but less effort may be put into these decisions because equipment isn't a big part of their business. Debt service and

DOI: 10.1201/9781003500315-15

maintenance costs exist for both contractors who own very little equipment, and for equipment-intensive contractors. The principles in this chapter apply to both equipment-intensive firms as well as those who use less equipment.

How Much to Own

The first step in controlling equipment costs is to control the amount of equipment owned. The decision to purchase new equipment is generally made for two reasons: either to replace aging equipment or for expansion. Both are certainly valid business reasons and must be considered carefully because the company is committing money to a business with limited assurances of future work. In replacing equipment, management should weigh:

- Is the new equipment more productive than what we have?
- If so by what margin?
- Is the difference worth the investment?
- Is the existing piece of equipment a maintenance headache?
- Will reconditioning get a few more years out of it?
- Is replacement more economical?
- Are the next two to five years of work a certainty?
- Is the marketplace growing or shrinking?
- Is the market likely to change soon in either direction?

Of course, no one can answer these questions with certainty, but asking them internally is often enlightening. The decision to purchase new equipment, which must produce profits over a period of years, otherwise it becomes a financial liability, is not an easy one. The decision to buy means taking on additional costs and increasing the necessity to get at least enough work to keep the equipment busy and justify the purchase. Too often we *want* to buy new and bigger equipment rather than *need* to buy it. When a replacement is needed, we tend to assume that bigger is better. Equipment-intensive contractors can fall into the trap of believing that more equipment is an indicator of success and set that as a goal. The only indicator and therefore goal is profit.

Reasons to Purchase

The decision to buy additional equipment for expansion is usually made for one of two reasons: New work is already contracted for and there is no owned equipment available to do work. Or a desire to have the equipment on hand in an expanding marketplace to accomplish the anticipated greater volume. If the marketplace is growing, it is reasonable to assume that the company will get their share of the growth. The problem with buying more equipment in advance of getting the work is that the company then *must* get more work to justify the investment and keep the equipment busy. The equipment is then 'running' the business instead of the other way around. Increasing owned equipment can create competing realities.

More equipment is required to expand, and more equipment forces expansion. The question that must be answered in an expanding market, is the growth permanent or temporary? Market growth driven by an increase in population is likely permanent while market growth because of the expansion of a flood control project may be temporary. There are too many failed construction businesses to count that were caused by expanding equipment and overhead to take advantage of growth that was only temporary. This is repeated regularly when a big highway bypass projects around a town or a one-time water or oil storage facility project comes to an end. Failed contractors regularly complain, *am I supposed to predict the future?* I trust that readers can answer that question. Being in business requires a lot more skills than just building the work.

Competitive Position

One of the difficulties presented by a growing market is that it can affect a company's competitive position. A firm's competitive position relative to their competitors is subject to change even in a stable market and more so in a growing market. At any time, a competitor's appetite for work or growth may evolve. A competitor may be going after work more aggressively than in the past, particularly during an expanding market. Market growth often attracts outside competitors who bid aggressively to get the first job and expand into your market. This may cause other local contractors to overreact by bidding even more aggressively. Keeping new equipment busy becomes a burden by forcing the organization to go after a greater amount of work, which can be particularly detrimental because they must bid at a lower markup to beat the increased competition.

Even if a company already has more work than their current equipment can handle, competition during an expanding market increases purchasing risk. When the existing work starts to finish up and equipment becomes available, the company may struggle to keep idle machinery busy. Increasing the inventory of equipment must be carefully thought through, not only as it relates to the work, but also as it relates to both the current and future market. The company's competitive position, and its resources to do the work at a profit must be carefully managed to fit the current and anticipated market.

There are times that current available work requiring more equipment may have to be passed up because the anticipated future market may not support the additional equipment. In similar circumstances some contractors have purchased the equipment and taken the work saying, *I'll cross that bridge when I get to it.* And when they got to it, there was no bridge. Successful contractors have loaded up with equipment in good times only to be forced into severe difficulties by the very same equipment when the marketplace went back to normal. Some have failed claiming the market caused their bankruptcy when the real cause was their failure to manage their business within the market. The equipment begins to run these contractors instead of the other way around.

Equipment Cost Rates

Calculating and accounting for the cost of owned equipment is ignored by some contractors or lost sight of by others. Some assume their accountants are taking care of that. To estimate work, a contractor must know exactly how much the required equipment is going to cost per unit time. To do that, the unit cost has to include all future anticipated maintenance expenses and the estimated replacement cost when the equipment is consumed at the end of its useful life. A contractor must have this information to price the work and to know whether ongoing jobs are profitable. The basic concept of unit costs for equipment is not complex, but determining it can be highly subjective. The primary objective of operating a business is to produce net income, which only results when more is received from a customer for services rendered than the total expense of producing the service.

When assets such as equipment are used in operations, they lose part of their service value, or the original value *is depreciated.* This depreciation expense is a portion of the actual cost of the asset that expires in each accounting period during the useful life of the equipment. This periodic cost requires no current cash outlay but is nevertheless a continuing expense for operating the business. There are a number of methods that may be used in calculating depreciation by Generally Accepted Accounting Principal (GAAP) rules, which are all based on the purchase cost of the asset. GAAP rules do not allow depreciation to be based on the estimated replacement cost of the equipment. This is in spite of the fact that when the equipment wears out, it must be replaced at its current cost at that time, which is always more than its purchase cost. This is a serious problem for the equipment-intensive contractor who has only recovered the original purchase amount through charging for depreciation.

Time and Usage

The cost of owning equipment is a function of both time and usage. Some equipment may be busy all the time under normal one-shift-per-day conditions, while other equipment may be used only periodically. A rock crusher, for example, typically works all year and is intended to operate all day, every day. If purchased new, the direct costs to the company during each month of operation, assuming mobilization is charged separately, are fuel, insurance, and regular maintenance. These costs are fairly easy to track monthly because they will be incurred during each month of operation. However, for major maintenance with spare parts required over time, an allowance must be included in the ongoing equipment costs. The allowance for maintenance and parts is an estimated cost that should be tracked and corrected occasionally to reflect what actually occurred.

This allowance should be treated as an actual (if not incurred) cost because at any time the crusher may need new belts and bearings, and the cost for these are not chargeable in the month incurred. The belts and bearings are consumed over many months, and this maintenance cost should be charged in a timely manner by estimating it in advance in an attempt to reflect reality. These very real costs must be charged as part of the usage costs to account for the true cost of equipment ownership, but because some of these costs are future expenses, they cannot be arrived at from current accounting records. They must be estimated.

Estimated regular and extraordinary maintenance costs are very real and should be included in the unit charges for the equipment owned because they are necessary to keep the equipment operable. If these costs are not charged to the unit costs, then they will come out of profits. The estimated cost should be applied to unit costs the same as incurred costs to ensure monthly costs are accurate. Estimated maintenance costs should be updated periodically (usually annually). If the equipment requires major overhauls such as engine or track replacement after two or three years, these costs have to be factored into the estimated maintenance and accounted for from the first month the equipment is put into service. If this is not done, the company will be overstating their real profit by not charging the wear and tear of the equipment to the jobs where it was incurred. Otherwise, it creates a false economy.

Replacement Costs

The inclusion of depreciation costs in the unit cost of the equipment does not cover replacement costs. The reason is that replacement cost at the end of the useful life of the equipment is always greater than the original cost of the equipment. The cause of the increase over time is simply *inflation.* For contractors to recover the *replacement* cost of a piece of equipment that now costs more than the purchase price, the inflated portion of the current cost must be included in the unit cost. This is in addition to the deprecation cost already included in the unit cost. The replacement cost is calculated in advance by estimating the useful life of the equipment and using that to estimate the appreciated replacement cost. This would be less any salvage value at the end of its useful life. The replacement cost is then spread over the estimated chargeable time during the equipment's useful life and added to its unit cost. In this calculation the number of hours, weeks, or months the equipment is anticipated to work per year is used. The replacement cost represents the cost to the company of using up or consuming the piece of equipment.

If an organization uses the purchase price and not replacement costs, as many do, they will not collect the increased replacement cost caused by inflation. A contractor once asked: *Do you really want me to charge my clients for next year's inflation when I'm only working for them this year?* My answer was: *No. You can always pay for it yourself.* If an equipment-intensive contractor charges only depreciation costs, they will, in effect, be paying for the privilege of being in business. They will be consuming equipment at rates calculated to repay the company only the amount they paid for the equipment. In which case, when the equipment is replaced after five years, the company will need more money than what was collected through the depreciation of the original price.

To be self-sustaining, an operating business must replenish itself from operations, which include the replacement cost of equipment consumed. Inflation is a cost of doing business and for an equipment-intensive contractor, this means charging replacement costs as part of the unit cost of the equipment. An economic analysis is required to determine the optimum useful life of a piece of equipment that will yield the maximum return on the investment. The time period that yields the maximum profit on an equipment investment may be considerably shorter than the operational life of the equipment. Over time the productivity of a piece of equipment deteriorates and repair costs increase from usage. There comes a time when the cost to operate the piece of equipment increases to the point where it is no longer economically practical. The equipment operates, but repair costs reduce its viability. Equipment costs include both ownership and operating costs. Operating costs for most equipment increase over time as the equipment wears out and requires more maintenance effort and additional costs to keep it in service.

Equipment Obsolescence

Equipment-intensive contractors have another exposure that is not very apparent and often not planned for. Which is equipment obsolescence. Companies incur a great deal of cost in replacing equipment as it wears out. Broken-down equipment delays jobs, hinders progress, and costs money. However, obsolescence can occur well before machinery reaches its useful life. The productivity of equipment dramatically affects profitability and is a big part of a contractor's competitive edge. Equipment productivity is critical to making a profit and to bidding and getting the work.

Oddly enough, perfectly good and relatively new equipment can become obsolete when a more productive piece of equipment becomes available. As newer and more productive equipment comes into the marketplace and competitors buy it, a contractor can be forced into equipment replacement earlier than planned just to remain competitive. Equipment obsolescence prior to useful life is a difficult issue because it is almost impossible to predict and consequently to plan for. It is therefore a risk of doing business for equipment-intensive contractors. It is difficult to compete with a firm that is more productive than you. To remain competitive, a contractor may be forced to invest in the more productive equipment solely because a competitor has already done so.

Equipment Obsolescence Case Study

A well-established Heating, ventilation, and air conditioning (HVAC) contractor with aging duct-making machinery was faced with some new and unexpected competition. A start-up contractor entered the market with the latest duct-making technology. The exceptional productivity of the new equipment allowed the start-up contractor to bid lower on every job of any size that came out during their first years in business. Eventually the well-established contractor began to run out of work. The established contractor realized that the only way to remain in business was to replace his outdated equipment with the more productive modern machinery the new competitor was using. Like most contractors, he did not reserve money for equipment replacement or, for that matter, even account for any replacement costs.

The company's current financial statements reflected a bad year because of the new competition, and the established contractor was unable to purchase the new equipment because they could not secure the financing. In fact, the new equipment was so expensive that their last five years' total profit wouldn't have paid for it. This established contractor had been very aware of his potential equipment obsolescence and understood the benefits of new technologies. He had been keeping abreast of the latest developments in his field such as computer-operated duct fabricating machinery, and more. In fact, he knew that eventually some or all of his equipment would need to be replaced. He simply never did anything about it because he felt that his equipment still had a lot of good years left in it.

With all this knowledge about modern equipment, he never thought of including replacement costs anywhere in his cost accounting. While ignoring the reality that at some point his equipment would require replacement, the contractor was enjoying what I refer to as *false profit*. That is by passing on to his customers the savings from not charging and accumulating equipment replacement costs. Basic economics requires that equipment-intensive contractors charge for and accumulate the replacement cost of the equipment they are slowly consuming by wear and tear and unanticipated obsolescence. Had this contractor accounted for replacement reserves, he would have realized at a minimum that his actual profits over the years were nowhere near what he thought they were. Additionally, had he considered obsolescence, he may have planned for continual upgrading of equipment or at least measured the risk of falling behind the latest technology. He should have quantified the risks and the potential cost of falling behind.

It is imperative to understand that competition is a strong force in any industry which includes all construction specialties and sub-specialties. The notion that an established contractor may not be expected to foresee the eventuality of a start-up company entering its marketplace is not the issue. New businesses with better ideas are a reality in any industry. In addition, existing competitors may gear up at any time to increase their market share. Out-of-town contractors are always on the lookout for new areas to expand into, particularly if existing competition appears weak or less productive. An equipment-intensive construction enterprise that does not concern themselves with the potential impact of obsolescence on their business is continuously at risk. All businesses must remain at least as productive as their industry average. Simply reserving the anticipated cost of *keeping up* is not enough because a company may not be able to gear up fast enough if their competitive balance shifts rapidly. It is imperative to invest in keeping up with local standards and advisable to keep up with national standards because unanticipated competition can originate from anywhere. The replacement cost of equipment, including costs due to inflation or obsolescence, will eventually become due whether or not it is reserved by contractors. Replacement cost is a cost of doing business that is incurred every day whether it is recognized or not.

Case Study: Replacement Costs

If an equipment operator decides to go into the dirt-moving business and buys a $100,000 bulldozer, how should he account for the ownership of this piece of equipment over the years? To simplify this example, let's say he buys it for cash from personal savings and that it will last for five years, at which time it will be worn out—no salvage value. There are a number of ways to account for depreciation, and he selects straight-line depreciation over five years, which is $20,000 depreciation a year. This new business recovers all other usage costs during the five years, including the $20,000 in depreciation. Where will it be financially in five years?

The business will no longer have their $100,000 because they spent it buying the bulldozer five years ago. They won't have the bulldozer because it is worn out and has no salvage value, and the contractor won't have a job because he doesn't have the piece of equipment to work with. What happened to the $100,000? It has been consumed by the business. Sure, they made profits during those years, and the depreciation allowed them to have $20,000 of the profit without corporate tax, but the $100,000 the contractor started with was after-tax dollars. The contractor spent the company's profits on salary and operating costs. The new bulldozer today costs more than the $100,000 that it cost five years ago. Let's say it costs $150,000. To stay in business, they need to borrow the $150,000. In this situation, the contractor is not only short the $100,000 that he had five years ago, but he is also in debt another $150,000 for the purchase of the new bulldozer. While the example is simplistic, it provides food for thought (Figure 14.1).

The replacement cost to be incurred five plus years in the future would need to be calculated and tax considerations would impact the calculations. Nevertheless, replacement costs will become due at some point and will almost always be more than the original purchase price, so one cannot rely on allowable tax depreciation alone to account for the cost of ownership of equipment. If the company buys equipment on credit and replaces it with credit, as most contractors do, the longer the company remains in business it will continuously go deeper into debt by at least the rate of inflation. In our example the equipment replacement cost won't be incurred for five years, which is not a problem unless you intend to be in business longer than that. The point is if this contractor had saved the depreciation, he still would have had to come up with or borrow another $50,000 to buy the replacement bulldozer. To remain in business, the company would have to accumulate and reserve $50,000 over the five years. The only source for the $50,000 would be profits, so actual profits would be reduced by that amount.

Even though the Internal Revenue Service (IRS) does not allow funds reserved for equipment replacement costs to be held tax-free, these funds are clearly a cost of doing business. While there has been some debate on this subject, it would be beneficial if the industry united

Figure 14.1 The unit cost charged for equipment needs to include the replacement cost.

and determined what the optimal method should be for accounting for *actual* equipment replacement costs. Ignoring the future replacement cost of equipment required to remain in business is operating in a false economy. Equipment replacement is a process that forces a company to go farther into debt over time, creating a serious long-term financial problem. The reality is that expenses incurred and due in the current accounting period are not much different than expenses caused in the current accounting period that will be incurred in the future and will become due in a subsequent accounting period. These future expenses are just harder to recognize and account for. In the example used here, the equipment replacement costs won't be incurred for five or more years, which is not a big problem unless you intend to be in business longer than that.

Idle Equipment

As equipment costs are incurred, they are charged weekly or monthly into a cost control system to give the contractor a realistic picture of whether a project is making is losing money in time to do something about it. Equipment is not an overhead cost any more than moving equipment back to the yard stops ongoing costs. Consider the heavy equipment that is not working. During downtime there is no fuel cost, and maintenance can be suspended, but the insurance cost goes on as does replacement cost accrual. Replacement is a function of usage, age, and obsolescence. It is often obsolescence that causes replacement, so the timing of replacement is not just usage. If a contractor believes that downtime will extend the useful life of the equipment, then they can adjust the replacement cost as long as they factor in an amount for deterioration from storage and non-usage. Deterioration can be a costly factor because most construction equipment wears better in use than out of use. Taking work at breakeven just to keep equipment busy is never justified except for survival.

The alternatives for idle equipment are not encouraging. To take work cheap or in other geographic locations just to keep equipment working is seldom worth it. To take highly competitive work just to break even is rarely justified. Liquidating some equipment is an alternative but must be considered in the context of the overall market, including new work anticipated. There is seldom a profit to be made in liquidating used construction equipment, although liquidation can reduce losses caused by the ongoing cost of idle equipment. Attempting to lease idle equipment to others is a favorable alternative, but this is usually difficult to do if there is a general slowdown in the market. If nothing can be done to mitigate the loss from idle equipment, the real costs of owning it must be charged to overhead, which serves as a reminder to try to find out where it might go next. It also tends to discourage project people from always asking for more equipment than they really need. When idle equipment is not chargeable, it is possible for all jobs to be showing a profit on paper while the firm loses money.

There is alternative thinking on this subject. Selling equipment because there is no work for it usually means that potential buyers of that equipment also have less work for their similar equipment. Therefore, the selling price for similar equipment is often considerably reduced. When the market recovers, if the same piece of equipment is needed because there is work for it again there is the strong possibility that other contractors may need similar equipment also. Therefore, few if any contractors will be selling that particular equipment because they will be using it. Therefore, the cost to replace that equipment may be considerably higher than the price it was sold for. There is some motivation to hang on to idle equipment for when the down market rebounds. However, idle equipment must still be stored, maintained, and ensured, and the equipment continues to depreciate in value over time. Idle equipment can also deteriorate over time. If the market downturn is short maintaining equipment works. If the length is unknown, a prudent move is to sell early. See section 'Alternatives'.

Cash Flow

Depreciation is a non-cash expense and thus contributes to cash flow. This portion of the cost of equipment ownership is not incurred with the equipment's usage; thus equipment-intensive contractors can have positive cash flow that can be mistaken for profit. Even if an organization accepts these concepts and accounts for all of the costs as described above, the company's cash flow will sometimes be greater than their real profit. If the depreciation, replacement, and extraordinary maintenance charges are not actually placed in reserve, then funds won't be there when needed. During slow periods with a lot of idle machinery, an equipment-intensive contractor could be showing losses on all jobs and still have positive cash flow, and able to weather the storm. If the company uses some or all of the funds that should have been reserved for equipment replacement, there isn't going to be enough money to maintain and replace the equipment when the time comes.

Because most equipment is purchased not with cash but on credit, the equipment is expected to be paid off from future work. For equipment purchased on credit, there is an increase in equipment expenses. Interest costs are added to unit costs similar to fuel and insurance. Since both interest and principal must be paid monthly, cash flow during slow periods is affected. During idle time or when losses occur for any reason, negative cash flow can quickly become an issue. Depending on the length of a slow period or a losing period, there may not be enough cash flow to make the equipment payments. The ownership of equipment is a two-edged sword. In busy times it magnifies profits and cash flow. In slow times it reduces profits and cash flow, sometimes dramatically.

Alternatives

Historically, equipment-intensive contractors had to own all the equipment they needed to perform their work because there were no options. Most of the larger highway contractors had their own mechanics, repair shops, and trucks and trailers to move the equipment. Leasing or renting equipment only became widely available about fifty years ago and provided an alternative to owning. For years there was a running debate over the advantages or disadvantages between owning or leasing. In a multi-year research project, I studied the issues, and the findings were conclusive. Equipment management is a critical aspect for equipment-intensive contractors and is a *science* not a *preference* or *opinion.*

Research indicated that owned and long-term leased equipment costs spread across its useful life vary, but on average are the same. Both are referred to here as owned and assumed to work one shift per day, five days a week unless described differently. Working longer hours or multiple shifts increases earnings but shortens usable life span. Some of the research findings were:

1 Owned and leased equipment is treated here as equivalent.
2 There are continuing ownership costs for both working and idle equipment.
3 Equipment working and chargeable is an income-producing asset.
4 Idle equipment is a liability with continuing costs.
5 Owned equipment is economically viable if billable 65% of the time.
6 It is cheaper to rent (not lease) equipment working less the 65% of the time.

For equipment that will be operated less than 65% of the time, it is more economical to rent as needed because owning is only economical if it works more than 65% of the time. Perhaps a more important aspect on this topic is that very often construction professionals use a less than ideal piece of equipment because the firm owns it. Experience indicates that this is a common occurrence in which an owned piece may accomplish 20 projects that 20 different pieces would have been selected for, if there was a choice. To keep the owned piece, busy the wrong equipment is used and never more productive. In multiple cases it was demonstrated that considerable money would have been saved (and therefore earned) by renting several different but correct pieces for each project rather than owning one piece of equipment and using it on multiple projects of varying appropriateness. Some need to be reminded that we are in the construction business not the equipment business.

The obvious challenge in purchasing new equipment with multiple years of useful life is knowing in advance if there will be work available to keep it busy (and chargeable) at least 65% of the time. Some argue that it is impossible to look that far ahead, which highlights the following point: When a contractor buys equipment expanding their capabilities, they need to understand that they are purchasing an asset and at the same time embracing a liability. The liability is that it is necessary to keep the equipment busy no matter what the cost because as per number 4: *Idle equipment is a liability with continuing costs*. If the market does not cooperate, they may have to take work cheap just to keep the equipment working.

The problem is hugely magnified when certain work activities require multiple pieces of equipment that are purchased as a group for a project. For example, a large earthmoving project may require multiple self-loading scrapers, one or more bulldozer machines to keep the haul road clear and dust-free. Finding another project for this group of equipment may be difficult. This demonstrates that even owning the equipment for a multi-year project that requires 100% usage presents the liability of finding work for this equipment at project completion for its remaining useful life or selling it. Managing equipment is complicated, and missteps have caused business failures. Readily available rental equipment has dramatically altered the options.

The ownership of equipment includes considerable ancillary costs of maintenance, insurance, transporting it to and from projects, and storing it when idle. The advent of lease and rental infrastructure has changed the equation. A contractor no longer has to also be in the equipment business. Some contractors remain in the equipment business, but many have come to realize that concentrating on producing the work is more profitable. Long-term leasing of equipment used 65% or more has the advantage of selecting the ideal pieces as needed, pickup and delivery, no cost of maintenance, and rapid replacement of a piece that breaks down. These things still cost money, but the equipment vendor does them efficiently and minimizes downtime. The contractor and crews can concentrate on producing the work while the vendor concentrates on keeping the equipment in shape. When it breaks down someone else repairs or replaces it at their expense.

Case Study: Idle Equipment

A nationwide earth moving contractor with over 1,000 pieces of heavy equipment, not counting rolling stock, owned and transported all their equipment and had mechanical shops with storage yards in multiple locations. They did all their own field maintenance and performed light and heavy repairs in their shops, including engine and track replacements (Figure 14.2). They were 100% self-sufficient with a massive staff devoted exclusively to equipment management. After a detailed study of their processes and costs, a new policy was put in place that disposed of all equipment that had a history of not being utilized at least 65% of the time. This resulted in a dramatic reduction in equipment costs and a significant increase in field productivity to the extent that after the first management decided to extend the study further.

The lifetime cost of owned equipment that worked more than 65% of the time was carefully calculated, and the average cost of each type was compared to historic lease costs. The lease costs were less across the board and significantly less in certain types of equipment. To alleviate

Figure 14.2 Field maintenance across multiple storage locations can be a significant cost to owning and maintaining a fleet of equipment.

concerns about availability, delivery, and efficiency, leased equipment was experimented with on projects in several areas of the country. The results exceeded bid profit in all cases. Over the next four years, as the company grew, it also converted to 100% leased and rented equipment, increasing profit percentage well over historic averages. These were difficult decisions because many people in the organization had a vested interest and belief in the manner in which the firm had done business for over 100 years. There is no implication that the old way was wrong. All business processes and industries evolve over time and as some say, *the only constant is change.*

Tools

Tools, including power tools, are similar to equipment in that they assist in and are critical to producing the work efficiently and profitably. Much that has been said about equipment in this chapter can be applied to tools on a smaller scale. The importance of tools is most obvious when they are not available. For example, when a crew of carpenters are building wooden forms for concrete and their power-saw breaks down, work stops. In some companies with *tool control system* paperwork and an explanation are required to get a replacement. In efficient companies that understand productivity, their projects have spares of every power tool on site. They don't have control systems because the systems always cost more to manage than the lost tools are worth. Productivity is primary and lost tools are meaningless. Lead and manage the workforce, not the tools. Laborers produce when they are supplied with everything they need and much of what they want.

Summary

Equipment is indispensable in producing the work, is critical to productivity, and can dramatically affect an organization's competitive position. Critical decisions include what equipment to use, how much to own, and whether it will pay for itself. Selecting the most efficient piece of equipment is almost always the best cost option and requires decisions about how much equipment to own, lease, or rent. It bears repeating from the chapter introduction: Equipment usage varies by type of work, but the decisions, costs, and management of equipment impact the efficiency and profitability of all contractors. Selecting equipment on both equipment-intensive projects and those with limited equipment requirements that turn out to be less productive results in reduced profits. The purpose of this chapter was an attempt to convince readers that equipment knowledge, choices, and usage are more critical to success than most construction professionals often realize.

15 Dangers of Known and Hidden Risks

Introduction

The willingness to take risks is a prerequisite for managing a construction enterprise, and it is unlikely that risk will ever be eliminated from the construction process. The ability to recognize the true nature of risk, assess its impact on an organization, and take steps to mitigate that impact is a fundamental skill set of successful industry professional contractors, owners, sureties, bankers, and designers. The contractor of the future needs to elevate *risk management* in their organization and embrace formal risk assessment processes. Becoming a risk expert is an immediate challenge for all construction professionals and should become the construction industry's newest discipline.

Owners have the easiest mitigation, as they can bond around their exposure and have a knowledgeable surety do the critical pre-qualification process, screening out ill-equipped enterprises and covering the costs if it does not work out. Sureties have an important role to play in the construction process and bring a unique perspective to contractor and subcontractor screening through the underwriting process along with financial capability when disaster strikes. Bankers can have the same protection as owners by requiring payment and performance bonds on the projects they finance. Designers can assist their clients with the selection of contracting methods and project participants while encouraging collaboration among all parties to the construction process.

Industry Risks

A business characterized by tight margins and periodic labor shortages that offers little room for error is obviously in a high-risk industry. Construction has experienced volatile market shifts in the past that have weakened some companies to the extent that they can have difficulty financing the growth that comes with the next market recovery. This in turn increases the potential for business failures and contract defaults. Now more than ever, owners, contractors, and designers need to increase risk awareness and risk protection. There is risk in any commercial transaction, but the construction industry has more than its share. Readers should be concerned that business failures in the construction industry are the second highest in the nation, and second only to restaurants. Much of the risk in the building process is poorly defined and often misunderstood with considerable ambiguity as to who is responsible for it. Designers avoid it, owners prefer to pass it along to others, and contractors absorb it.

Today's construction risk environment is different than it was three decades ago when industry risk factors were defined as changes in project size, type, geographic area, key personnel and/or managerial maturity. These were documented in the author's first book *Construction Contractors' Survival Guide* (Wiley) and updated in *The Business of Construction Contracting* (Wiley). There have been developments in the current construction environment that maintain and, in some cases,

DOI: 10.1201/9781003500315-16

magnify these business risks. The excuses are increased by the fact that most contractors have little or no fear of risk. Four decades of research into the causes of construction business failure uncovered a noteworthy truth: The assumption of risk is part of every successful contractor's DNA. It is not uncommon to hear a contractor say, or demonstrate by their actions, very little fear of risk.

Risk Categories

There are business risks common to every industry and additional risks unique to specific industries. Some are easily recognizable, others more difficult to detect, and the magnitude of each varies by industry. Industries that operate in structured environments like manufacturing or healthcare have managed to control much of their inherent risk with the introduction of standardized procedures. Industries in less structured environments like construction and mining are exposed to greater risks, which are much more difficult to control. The amount of control over the production environment is directly proportional to the exposure to risk. The high business failure rate in the construction industry verifies that every category and type of contracting is exposed to serious business and financial risks. The following are some of the major risk categories in the construction industry.

- Growth Risk—Growth is the primary cause of construction business failure and yet remains the primary goal of most construction enterprises. Growth is clearly not perceived as a serious business risk that it is. Research confirms that when a construction company expands in size, it requires careful management decisions to reduce the risks inherent in the change. A rapid increase in sales in and of itself increases the potential for financial distress.
- Subcontractor Risk—The multiple subcontractors on each project are each a fundamental part of the project's success, and therefore the success of each is dependent on the performance of the other. It only takes one subcontractor's misstep, mistake, or slowdown to disrupt the entire project progress and to diminish or dismiss profit potential for all. The sheer number of subcontractors and suppliers on a project magnifies the risk for each.
- Capital Risk—Both general and subcontractors rely on and survive on secured equipment loans and unsecured working capital lines of credit. In many cases they also depend on surety credit for bid, payment, and performance bonds. These financial elements are vital to the success of each construction enterprise, and any disruption for any reason can put the company and all their projects in jeopardy. Financial instability is a risk multiplier.
- Commodity Risk—Construction is a custom service and product that many project owners or designers believe is a commodity. This causes some owners and designers to think all construction companies perform alike and as such, they become less discriminating about contractor selection, thinking price is all that matters. When buyers or designers of construction services believe the product is a commodity, they expect to pay less. This is one of reasons profits have fallen from historic norms.
- Contract Risk—Over several decades in this highly competitive market the contract terms concerning the responsibilities of each party have been mutating. There has been a shift of more and more project risks moving towards the contractors and away from the project owners and designers. At the same time, case law (court rulings concerning construction disputes) seems to be further clouding the question of which party is responsible for what as the new contract terms are explicated.
- Change Risk—Business processes and beliefs change and evolve over time. Some claim that the only thing certain is change. Well-intentioned but ill-informed changes by construction professionals can and have caused business distress. For example, growth changes increase the

amount of capital required. Changes in type of work or in key personnel can reduce the time and attention management has available for other critical decisions. Caution is required because multiple changes have stretched existing expertise to the breaking point. Change must be managed.

- Technology Risks—The increasingly rapid advancements in the construction industry's field and office technology have introduced potential risks. Selecting and implementing new technology consumes management time and energy and, if the selected technology does not perform as expected, it can be costly. Avoiding new technology is not a solution, so managing the introduction of new technology has become an additionally required business skill.
- Financing Risk—Innovative private and public funding methods present risk to all construction professionals simply because they are new and there is limited experience with them. Changes in the financial industry serving the construction industry continue to evolve. These demand new knowledge and management attention that expose construction professionals to increased and sometimes unrecognized risks. Another business skill is required.

Risk Assessment

Management decisions will determine whether an organization will succeed or fail in the ever-changing construction business environment. The decision-making process relies on knowledge and beliefs that must be regularly re-examined and updated as the business environment evolves. Business beliefs that were appropriate in the past may not be adequate in what some refer to as the ever-changing *new normal*. Some unexamined beliefs in place for a long time may no longer be valid such as that growth is always good, having some unprofitable work is unavoidable, and past success implies future success. These beliefs need to be reevaluated because they are no longer true, or may never have been true, and should not be embraced by the *successful construction professional of the future*. (The term *successful construction professional of the future* describes construction professionals who react quickly to evolving market conditions). Forty years of accumulated knowledge as a contractor, workout specialist, consultant, and research professor verify that:

- Construction contracts include *risk assumption*.
- Risk-taking is embedded in a contractor's DNA.
- Each contract signing is like another first day in business.

Every new project includes voluntarily assuming risks not fully defined in the contract documents. Each new project is a highly complex endeavor, worked out over a relatively long period of time, whose success or failure is affected by, but not limited to, weather conditions, labor problems, and inflation. It is also subject to unexpected increases in interest rates and cost of equipment. As if that were not enough, there are exposures to changes in field conditions, growing or shrinking markets, or simply bad luck. These make the measurement of project risk in advance enormously difficult, to the degree of almost impossible. This challenge requires an objective appraisal of the capability, capacity, strengths, and weaknesses of your construction organization. Also required is a detailed understanding of the parameters of the potential project, and of the current construction environment and conditions. Enlightened construction professionals expand their *knowledge resources* and test their beliefs and perspectives with others. They seek input and information from their trusted competitors, successful subcontractors, internal and external accountants, and their surety and/or insurance professionals.

Hidden Risk

Business and project risks are often not apparent and may be disguised in many forms. For example, top-line growth may appear to be an easy avenue to success but can also lead to failure. More tightly drawn contracts pushing any-and-all risk to others may appear to be a needed protection but can create more problems than they solve. An expanding market that looks like an opportunity can also be a minefield. The contractor of the future needs to learn an entirely new skill set to recognize risks hidden in the market, hidden in their management decisions, and hidden in the economic climate. Too often what looks like good news has the potential to be hazardous. Some risks cannot be eliminated but can be controlled. To control risk, it must be identified, measured, and thoroughly understood. This is complicated by the various parties to the construction process seeing their respective roles in addressing risk differently. Many construction risks are attributed to more than one entity, making identification difficult, elimination impossible, and control or mitigation of the risk the only viable alternative. The challenge for contractors, owners, designers, sureties, and bankers is to:

- Recognize and identify specific risks in advance
- Assess and quantify each risk's magnitude and exposure
- Mitigate and manage the impact and cost of each risk

To accomplish this requires specialized industry knowledge, intelligence, and experience. It is also necessary to have access to hard-to-find, and sometimes very confidential information. Contractors have a reliable source for risk information, measurement, and assistance in their surety company, which has extensive experience and data on all types of construction in various locations. Sureties understand financial risk and should not be criticized for what some contractors complain is restricting their growth. Contractors often want larger projects or extensions of bond totals than their surety is willing to provide. The contractor's position is usually that this project will generate *huge* profits. The surety's position is usually if this *huge* project that is larger than anything you have built before does not go as expected it can put you out of business. It is a matter of perspective. One is driven by optimism, and the other is driven by practical business sense, which includes *risk control* (Figure 15.1).

Biggest Risk

In the construction business, past success is not an indicator of future success. Research into the causes of construction business failures indicates that every change in a successful organization like growth creates a period of risk despite any previous success with growth. When a construction organization increases in size by more than 15%, it is no longer the same company that it was before growth and may not be successful if it maintains pre-growth management practices. A construction enterprise growing at even the safe rate of 15% per year will expand rapidly if the expansion is continuous because it compounds quickly. At a 15% rate of growth, a company doubles in size in five years and triples in eight. This continuous growth is significant because few construction markets grow continuously by anywhere near 15% annually.

Growth at a rate greater than the rate the current market is growing makes it almost impossible to maintain profitability. The successful construction enterprise of the future will be organized to be market-driven and not volume-driven. It will strive for carefully planned growth but be prepared to level off or fall back in volume if the marketplace tightens or shrinks. It will use its markup flexibly as a competitive tool but never be forced to take low-profit work to maintain sales. It would

Figure 15.1 Many construction risks are hidden, requiring deliberate analysis and proactive management to uncover and control them.

have some flexible overhead built into the organization that could be cut immediately when not required and would not hesitate to cut permanent overhead if downsizing becomes necessary. The successful contractor of the future would never hesitate to reduce in size to survive. (For a detailed explanation of benefits and how to use Flexible Overhead, see chapter four of *The Secrets to Construction Business Success*, by T. Schleifer and M. El Asmar, Routledge, NY, London, 2022.)

Risk Management

Historically the discipline of *risk management* was considered by many to be exclusively an insurance issue whereby insurance coverage was purchased to cover known or anticipated risks. However, there are numerous significant business risks that cannot be covered by insurance. There are hidden risks in market expansion such as skilled labor shortage, limited access to capital, and the ability of management to keep up with a rapidly changing business environment. These hidden risks have rendered traditional insurance tools and risk management approaches inadequate. The successful contractor of the future will develop internal risk management policies, procedures, and protocols throughout the organization. Some enlightened firms will engage or employ risk professionals.

Construction is a complicated business that should only be attempted by contractors, subcontractors, and senior construction professionals with strong capabilities in risk recognition, management, and mitigation. The nature of construction risks is changing every day, increasing the potential for business and project failures. Unexpected swings in the marketplace force construction firms to struggle with fluctuating sales. The shortage of skilled labor has already caused some to downsize or spread their skilled labor more widely. When and if new people can be found, they will often have limited experience and be unfamiliar with an organization's methods and systems, creating productivity issues. Rapid growth often surprises contractors who did not foresee the cost requirements to discover they have inadequate access to capital. Some also face aggressive buyers who view construction as a commodity.

Market expansion has included some unexpected and unrecognized risks that render outdated beliefs a risky mindset. The construction industry has not demonstrated proficiency at recognizing, managing, and mitigating hidden risks. And growth escalates the potential for business and project missteps. Owners can manage their risk by bonding prime contractors, who in turn manage their risk by bonding their subcontractors. The successful contractor of the future will establish formal risk assessment processes and protocols that fit the newly recognized cyclical construction market. The enlightened project owner of the future may choose to transfer their risk by relying more heavily on sureties to prequalify contractors and provide financial protection against defaults.

Evolving Risks

Even in good markets, bad construction enterprises can fail. Financial difficulties in good markets are unexpected, but unforeseen business or market changes cause reactions as well as overreactions. There is little agreement among contractors regarding how to react to unexpected business or market conditions. A recent example demonstrates the complexity. An underperforming general contractor decided that what he needed to do to reverse weakening profits was to grow bigger because the marketplace he served had expanded. Over several years a few of the contractor's competitors grew substantially or merged with larger firms. At the same time, quite a few of the contractor's steady customers had been acquired by national firms and began to contract with larger, regional, or national contractors for their construction and maintenance needs. The market the contractor dealt with was changing, and work available to his size organization was shrinking.

This contractor's limited resources made growth difficult, but his argument was that he did not put his entire life into building this company to get smaller now. The growth he undertook may or may not have been profitable because before he could find out, his cash flow dried. He was unaware that growth eats cash and therefore was unaware of the cash flow demand he would experience. What this contractor did not understand was that for the last decade or more his industry was experiencing *consolidation* that was impacting contractors, owners, and designers. All of whom were being forced to adjust to the changes they did not understand. Many US industries have consolidated, so the process is predictable. Typically, big firms will get bigger, while midsize firms will merge, be acquired, or otherwise cease to exist. Small firms will be least affected. Contractors have been almost universally resistant to the data that indicates that the industry has begun to consolidate. It appears that most will not react until it is certain; however, the unwelcome news is irrefutable.

Industry Consolidation

Construction processes are rapidly becoming standardized as technology simplifies the work. Technology can take a complex and time-consuming process and streamline it into an efficient less cumbersome operation. For example, not that long ago tradesmen built wooden forms from scratch to produce complex concrete placement. This work was limited to highly skilled tradesmen with experience at producing concrete forms. But the process was revolutionized by patented form systems that require much less skill to install—almost to the degree that anyone can do it.

Standardization modifies an industry and encourages new entrants. Economies of scale help lower costs using standardized processes, but the time and costs to purchase new systems and learn to use them favors large companies. Most small and mid-size construction companies lack the resources to reduce costs and schedules as fast as customers demand. Larger enterprises can afford the resources required to reduce costs and schedules, and with the economies of scale, they can benefit from process improvements that foster consolidation.

Historically construction was a "can-do" industry of brawn and might, with a tradition of building projects with basic equipment and limited technical support. A lot of projects were built with a backhoe and a box full of tools. It was claimed in the past that all you needed to be a contractor was pickup truck, a cast-iron stomach, a forgiving wife and a bad temper. The truth is that historically, success in the construction business demanded a rare combination of talents, including ability to:

- Muster the necessary resources
- Put an accurate price on the work in advance
- Manage labor, subcontractors, vendors, and designers
- Tolerate a high degree of risk

All of these talents are acquired through a long and arduous process.

Contractors with these capabilities had great success if their estimated costs for individual items of work were closely guarded secrets. This was aided by the fact that the production of complex projects remained a mystery. But all of that has changed. New technologies and wide access to information have demystified the construction industry processes of estimating, organizing, and producing the work. What also changed was that project owners and designers became knowledgeable about construction processes and the underlying costs, which had been a closely guarded secret. Construction, which was historically a complex custom effort with double-digit profit margins to compensate for inefficient processes, is now becoming almost a commodity. Costs and production methods have become widely known. This is resulting in contractors today working harder for far less profit as margins continue to shrink.

Despite an exceptionally good economy, construction margins have slipped into the mid and low single digits and even lower for larger contractors. Margins have been *structurally* altered and will not likely ever return to historic heights. Therefore, as construction became increasingly standardized, efficiency and productivity became the major differentiators among competitors. Some organizations did not become more efficient complaining that they couldn't afford to upgrade their people because they were too busy for training. Low margins restrict resources, making it harder for small and mid-size businesses to invest in increased efficiency. Larger companies suffer the same low margins, but their size enables them to invest in productivity improvements to satisfy the increasing demand for efficiency and the lower costs being demanded by their clients.

Continuous improvement is a cost of doing business and is critical for survival in today's construction market. Small and mid-size firms may be able to accomplish large-scale efficiencies through cooperation, but their window of opportunity will be brief. Some small contractors will fare better than mid-size firms, because there will always be a place for niche players. Specialty contractors will enjoy relatively higher margins until consolidation moves into full throttle. Consolidation will then begin to encompass each of the subdisciplines, which will lead to even more specialization. Profitably well-managed subcontracting enterprises will survive by merging with or becoming part of larger organizations. Small, well-managed general and subcontractors will succeed by serving niche markets.

Safe Growth

Rate of growth for construction companies is always a challenge including the potential risk of growing too fast. The good news in a growing market is that there is plenty of work. The bad news in a growing market is the same, there is plenty of work. Growing markets and growth are a two-edge sword. Our industry always seems to move between not having enough work to not having enough labor to do the work. There is seldom a period when everyone says there is just the right amount of everything. Construction is a fragmented industry with almost a million and a half construction businesses, all with their own opinion about how the industry is doing. Most contractor opinions about the health of the industry seem be somewhere between short-term projections and guesses. Pessimism seems to prevail because during both good years and bad, contractor opinion polls always predicted a slowing market. These consensus polls seem to be more speculation than research because other market indicators are usually quite different.

Inaccurate forecasts have a negative impact because erroneous projections for a weak market cause contractors to price more aggressively than more accurate positive forecasts would. The industry often suffers more from depressed margins, some of which are influenced by pessimism more than reality. Reduced profits are to be expected in a down market, but pessimistic forecasts have the same impact. Taking on low margin work can cause financial distress or worse. Negative forecasting unnecessarily reduces profits in down markets, but it does the same in a growth market by minimizing the potential advances. During recoveries some contractors enjoying the rebound capture more than their share of the market and then falter from the overload. The casualties of a robust construction market suggest financial distress increases in direct proportion to market growth as well as a market reduction. The reason is that it is extremely difficult for a construction enterprise to calculate just how much it can effectively perform and finance. Worse yet, very few construction professionals understand that every organization has a very real growth limit. Unfortunately, the surest way to discover that limit is to exceed it. Adding to this exposure is that industry record-keeping can obscure problems for a long time to the extent that performance and finance limits may not be obvious.

Even profitable work puts a strain on cash flow. And few construction organizations can gear up quickly enough or solidly enough to hold profit margins during rapid growth. This is particularly true during a difficult labor market. A volume driven industry is inclined to excessive growth and business risks in construction are amplified during growth. There is high risk in the best of times which is amplified during expansion which is when management is stretched already. Contractors should not underestimate the increased risk from growth in the closely held construction enterprise. Overhead costs are difficult enough to control when companies are not growing, but in a growing organization just managing overhead is a hazardous problem. Because organizations cannot add a half-person or a half-piece of equipment, they are forced to incur overhead costs during growth in larger amounts than they would like. This creates losses until the company grows into the overhead. The problem is magnified when lagging profits create an absolute necessity to increase sales to cover the increased overhead. A vicious cycle putting the company in double jeopardy.

Growth Issues

As an organization attempts to increase market share, prices suffer because it is necessary to make at least temporary price concessions to take market share away from competitors. Even with no conscious decision to lower prices to capture added volume, it always seems to occur. When selling prices suffer, they usually suffer for all the new work, not just the occasional aggressive pricing. When margins suffer, the company ends up needing more sales than originally planned. This usually leads to a downward profit spiral because when an organization gets stretched, there is never time for anyone to see the problem developing. The chain of events when additional growth requires more overheads that creates the need for more sales has been the cause of numerous construction business failures. Rapid growth also puts a strain on a company's key people and systems, and sustained growth doesn't allow for a reasonable training period for new employees. Additionally continuous growth doesn't give an organization a chance to test new people or systems before the next new batch of people and systems are added.

If performance deteriorates because of growth, it will only be discovered after the additional volume and people are taken on. Corrective measures are more difficult with new people and systems being stretched while overworked managers are trying to cope with the largest volume the enterprise has ever handled. Some companies don't recover from this scenario. Too many organizations pursue growth without measuring performance until it is too late. It bears repeating that years of research indicate that growth for a construction enterprise of more than 15% annually should be considered substantial and adversely affects business risk. Sustained growth over more than a couple of years compounds quickly. At 15% a company doubles its size in five years and triples in seven. At 25% it doubles in three years and triples in five. And at 50% a company doubles in twenty months. Growth requires more resources in the way of people, systems, and money. Success is measured in an organization's ability to find the necessary qualified people, train them, put appropriate systems in place before expansion, and finance the increased costs.

The rate and the pace of growth obviously affect an organization's ability to bring adequate resources to bear on the new work. The alternative is to demand more from existing resources. However, few construction organizations are known for having underutilized resources or bench strength. As volume increases, an expanded company is untested as an organizational unit. The only reasonable test is for the new organization to operate profitably and smoothly for a minimum of a year. Sustained growth creates a situation in which, if the test proves unsatisfactory, new growth has already been added during the test year. The organization is then facing a second bad year before it can roll back to its proven size and proven team. For many it is too late to retreat and recover.

Limits of Expansion

Determining the limits of expansion for a construction organization is complicated. In fact, some highly respected management specialists don't believe there is a limit. A good number of construction companies that were well-known names in the industry have failed during meteoric growth, suggesting that limiting restrictive factors exist. Although critics may point to other reasons for failures during growth, the reality is that rapid growth, in and of itself, is dangerous. It is not always fatal but is always risky. Fundamental financial constraints limit healthy and sustainable growth. The management of growth requires careful balancing of sales objectives with the firm's operating efficiency and financial resources. It is critical to determine what sales growth rate is consistent with the current realities of the company and the marketplace.

Companies have limits in abilities, resources, and capital. Each organization is capable of doing just so much. During periods of rapid growth, closely held construction companies are so changed that they, in effect, become new untested organizations. This occurs at a time when they are producing a lot more work. The prior organization that was so successful is gone forever. For growth to be successful, quality production must be preserved, but growth dilutes quality unless expertly managed. However, during growth, management is stretched and expanding. It takes more time to increase management and supervision and to maintain quality than it takes to capture more work. Most companies decide to increase management and supervision only after additional work is on hand, not before so quality is challenged. Growth just for growth's sake is risky in any business, but growth in the construction business is far more complicated than is commonly believed.

Incremental Growth

Incremental growth instead of sustained growth may seem unnatural, but it is the best way to control the inherent risk in growth beyond 15%. Incremental growth allows cycles of growth and testing, then growth and testing again. In this manner an organization can reevaluate goals and recover after a poor test. The alternative is to pursue constant growth until incurring negative results from which the company may not be able to recover. This is prudent risk control. In sustained growth, a company grows beyond its people and systems so often that it never has the same organization long enough to truly test it. The company is then functioning at constant risk with an ever-changing team. In some cases, failure is just a matter of time. Incremental growth is prudent business management that requires contractors to grow with care, test as they go, and be prepared to withdraw from bad decisions.

Growth eats cash primarily because construction enterprises put the work in place and wait for their money. If a company is continually putting more work in place, it will eventually run out of cash and credit. When more work is available, it is difficult to turn it away but also difficult to accurately project how much work a firm can effectively perform and finance. The consideration is if the annual growth exceeds 15%, it will increase risk to the extent that a contractor needs to

Figure 15.2 Growing too fast and the resulting strain on a company's resources is similar to an overloaded truck attempting to climb a long steep slope.

think twice and plan well. How will the additional work be produced, and where will the internal or external financing come from to support it. Construction professionals should understand that it is extremely difficult to forecast how much an organization can effectively perform and finance. Every organization has a limit. Consider carefully how additional work will affect the organization and approach cautiously. If you are concerned, consider the extent of your present risk and the wisdom of adding to it.

Managing a closely held construction company can be compared to driving a truck up a hill (Figure 15.2). The steeper the grade, the more strain on the truck, engine, suspension, and drivetrain. More work for an organization places more strain on management, employees, systems, and finances. A truck starting up a hill from a level roadway finds it easier than starting already on a hill and increasing the slope. Separate short climbs are much easier than sustaining continuous uphill progress. Trucks attempting a very long and steep hill often slow to crawl, labor to gain forward progress, and some actually stop. When a construction organization embarks on a steep climb at a growth rate of more than 15%, it will always experience a strain on its resources. During periods of continuous growth, the strain is sustained and magnified, sometimes to the breaking point. A truck advances much more efficiently up a series of modest grades than up a long steep sustained climb. Managing strain on company resources increases efficiency, profitability, and risk control. Subjecting resources to severe or continuous stress encourages inefficiency, deterioration, and potential decline.

Summary

Risk is inseparable from the construction business, and while it can never be eliminated, it can be identified and effectively managed. The contractor of the future must recognize that every period of growth introduces new risks, many of them hidden and often more dangerous than those that are obvious. What was once a business of intuition and bravado has now become a discipline that demands structured risk management and the ability to adapt to changing markets and technologies.

Sustainable success in construction is not built on growth for its own sake, but on profitable work delivered within the financial and managerial capacity of the organization. Incremental growth of no more than 15% annually, supported by formal risk assessment processes, offers the best path forward. Growth beyond 50% a year carries extremely high risk. Contractors should never lose sight of the primary goal of a construction enterprise: profit, not volume.

Growth in any construction organization increases risk. Continuous growth is extremely risky; therefore, incremental growth is recommended. There are very real limits to a construction organization's rate of growth. Growth at the rate of 15% a year or less is at minimal risk. Over 15% annual growth risk increases proportionally as the rate increases. Growth above 50% a year has extremely high risk. The primary goal of a construction organization is profit not growth.

16 Industry Market Cycles

Introduction

The construction market in the United States is cyclical. The market has been cyclical since the 1950s, but that was unknown until I discovered and announced it more than 40 years later. As it turns out construction has always had cycles caused by supply and demand, war, weather, and economic conditions. The first major construction projects that there is any record of were cyclical. Irrigation canals used to divert water from the Nile River in ancient Egypt were built during the dry season between the annual flooding of the river. These were followed by the building of the great pyramids, and the construction industry was born. An industry that produced the built environment through to the present day. A detailed study of the industry in the United States from the end of WWII to the present day indicates there was continuous growth interrupted by short cycles of negative growth. Measuring the US construction market in dollars, it grew from the 1950s through today. It grew in cycles of 10–12 years interrupted by one- to several-year periods of market declines.

This cyclical 'pattern' was discovered and documented about two decades ago after a multiyear research project by the author. The downturns occurred for various reasons generally resulting from US economic slowdowns, the worst of which were referred to as recessions. The research determined that the construction market downturns were a result of reverses in the general US economy. However, the downturns actually followed these reversals in the general economy by twelve to eighteen months. The reason for this *lag time* is that at the onset of an economic downturn, construction funding weakens, but projects already funded continue to move forward. As a result, construction being put in place measured in dollars does not slow down for another twelve to eighteen months after the onset of an economic downturn. The same delay or lag also occurs after the US economy recovers. The reason being that upon recovery construction projects begin again to be funded, after which they enter into the design phase. These two phases can take twelve to eighteen months before the projects are ready to begin the construction phase.

The announcement of the research findings about the cyclical nature of the construction market and the timing and average length of market downturn seemed to be of little interest to contractors, designers, or project owners. Either most of them never heard about the findings, they didn't believe them, or thought they didn't matter. The significance of these findings for an industry dedicated to growth is that *continuous growth while maintaining profitability is literally impossible* in a cyclical market. One reason is that most construction company overhead are *permanent fixed costs*. As a construction organization increases the amount of work they put in place (growth), it requires an increase in overhead costs to manage and produce the work. Unlike most field costs at a construction site, which can be increased or reduced depending on the amount of work, overhead

DOI: 10.1201/9781003500315-17

expenses are primarily permanent costs. Costs of office personnel, facilities, and equipment are permanent costs not easily suspended in the short term.

Overhead

Overhead expenses for construction businesses today are primarily, if not 100% *permanent* costs, and these expenses are not easily terminated in the short term. Perhaps more importantly, most contractors have no intention of cutting overhead when their workload falls off. The typical reaction to a market downturn is to reduce profit margins, even to breakeven, in order to capture work and maintain the organization's structure for when the market rebounds. Because overhead is never considered temporary, cutting (permanent) overhead is considered *dismantling* the organization. Contractors generally believe that they have a substantial non-recoverable investment in their overhead personnel that they can't afford to lose under any circumstances. Once laid off, these people are lost to the organization because they are not going to be sitting around waiting to be rehired a year or two later. Owned or leased office space costs are almost impossible to suspend, and there are few buyers for construction equipment during a construction market downturn.

The inability to reduce overhead in a market downturn is not perceived as an industry problem because market downturns are considered rare. Because construction market downturns in recent history have been separated by 10–12 years or more, most construction professionals have experienced only one since they joined the ranks of top management. Experiencing two downturns while holding a leadership position would be rare, so the longstanding industry conviction that construction market slowdowns are very rare persists. This will continue until there is some acceptance of the proven reality that the construction market is cyclical and that downturns occur at measurable and regular intervals. The reason the discovery of the cyclical nature of the market is so crucial is that it demonstrates the absurdity of the belief that continuous profitable growth is even possible. *Permanent overhead* and *cyclical industry* are mutually exclusive. Something has to change. Considering that the construction industry has no influence over the economic cycles in the US economy, contractors must adapt. The solution is *Flexible overhead*, which is described later in this chapter.

Market Downturns

The research on cyclical markets clarifies the recognition and identification of market slowdowns that lead to increased competition. These provide a warning of the pending change in market conditions. The following list of events was consistent in each of the downturns since WWII.

- The average number of bidders on public and private projects dramatically increases immediately after the first sign of a reduction in construction funding. The longer the economic downturn, the greater the number of bidders. In one cycle the average number of bidders increased over time by a factor of four.
- The unit cost of construction of all types reduced across the country and continued dropping until the construction market recovered. The timing of recovery of unit costs in some downturns varied by area.
- The construction company failure rate begins to increase slightly about a year after the start of a market downturn and continues to increase for years after a downturn. The increase in failures is roughly proportional to the length of the downturn.
- Profit information is not published for at least a year after recovery, and some of the results can be delayed for several years. The averages captured are informative but may be understated

because of timing. Profits begin to reduce immediately upon the start of a downturn and suffer long after a lengthy recovery. The longer the downturn, the greater the obsession of contractors to capture the limited amount of work available.

- A down market tends to weed out underperforming or poorly financed firms. Losses are also always underreported as vulnerable and struggling construction firms are extremely hesitant to share their financial information.

An interesting finding is that mergers and acquisitions increase noticeably three to six years after each market recovery following a lengthy downturn.

Prospering in Cyclical Markets

Readers now know that market downturns occur at regular intervals and are usually short relative to the length of growth markets. When a construction company's backlog falls off for any reason, the pressure is on, and it feels like a recession regardless of the state of the economy. Prospering in cyclical markets, similar to surviving a *recession*, starts with understanding in advance what will happen in the marketplace as soon as a market softens. The results are totally predictable and have occurred without fail in every construction industry downcycle for the last 75 years. When there are fewer projects in any market, competition intensifies, and prices and potential profits diminish. In a shrinking market it would be ideal for each contractor to accept proportionately less work so that the market share of each business would be maintained. However, this has never happened.

There is a strong tendency in our industry to resist any reduction in sales and to fight vigorously for the fewer available projects during a downturn, which drives down prices for everyone. Trying to maintain sales in a declining market is in effect, the same as an attempt to increase market share. Increases in market share are universally *bought* at a cost. Contractors attempting to resist a reduction in their sales by taking on *cheap work* are dramatically increasing their business risk in an already difficult circumstance. An alternative is to *cooperate* with the market, which would mean accepting no growth and voluntarily downsizing their organization to align with the current market realities. This is the only appropriate reaction to the risks imposed by a change in the economic environment that is totally outside of your control. Taking cheap work in an attempt to maintain sales is not a viable alternative (Figure 16.1).

Downsizing

You cannot control the market, but you can control your response to it. There is potential for profit during a declining market, but profit is much more difficult to achieve than in a growing market, and it requires some ingenuity. It is every contractor's responsibility to react to and manage risks in both good markets and bad and to remember that profit is the prime objective—not growth in sales. This responsibility sometimes includes taking the difficult step of *reducing* sales in order to downsize and protect the business to remain profitable. Downsizing includes a simultaneous reduction in overhead as volume is reduced. There is a natural inclination to delay any reactions while waiting to see if the market improves. This delay is prompted by optimism, which is common among contractors. However, controlling the risk of overexposure in a declining market mandate that defensive steps of downsizing be taken sooner rather than later. In a market downturn the primary objective is to avoid any exposure to loss potential. Risk control never includes *speculating or betting* on market recovery.

To avoid any reductions in sales, contractors often argue that they can't operate with 10% or 20% less work, claiming there is minimum or a *drop-dead* volume, which is some amount they

Figure 16.1 Market downturns leave contractors competing for fewer projects, often resulting in idle crews and unused equipment.

feel is required to remain viable. However, if a contractor can profitably grow their business from five million to ten million or 100 million to 200 million on their way to 15 million or 300 million, they should be able to profitably downsize from 15 million to 10 million or 300 million to 200 million. Resistance is a refusal to give up hard-earned volume. If a contractor must go back to one of those reduced volumes to remain profitable, they will need to reduce and reconfigure their current overhead to what it looked like when they were profitable at the smaller size.

To accomplish this includes reducing equipment resources by selling or mothballing if selling is not practical. Contractors are reluctant to do any of this because it seems like dismantling an organization they built; so they risk losing it all while hoping for a short downturn. The inclination to hold on to people and equipment in order to be prepared when the market returns is strong. However, hoping for a short downturn presents serious risk that resources will be retained at great expense only to be let go at a future date. Or worse, the drain on the organization may make it difficult or impossible to finance recovery when the market rebounds.

There are numerous motivations to resist downsizing because it involves the *demotion* of employees in hard-earned positions. When a foreman becomes a superintendent, they usually don't want to go back to doing the actual work of managing the day-to-day people in the field, and they've already trained their replacement. In the office, project managers who used to both bid the work and then manage it when they won a job don't appreciate going backwards in their careers. To reduce overhead by scaling down the company means some people will be doing what they used to when the company was smaller. These are difficult decisions for company owners and non-owners alike to accept. However, survival of the company is simply more critical than these concerns.

Reducing overhead expenses is a thankless task even when there is no other choice. The starting point is cutting back nonessential costs such as subscriptions, bonuses, travel and entertainment, etc. This has a secondary benefit of signaling to employees that there is a new attitude towards expenses. Unfortunately, these non-essential expenses never amount to enough to really matter. The largest by far of all overhead costs are the salaries of management and administrative personnel. These cutbacks are also the most resisted by managers, who are usually against any overhead reduction for any reason. Refusal to cut overhead often results in business failures. These types of failures are directly proportional to the length of the market downturn.

Depressed markets always return to growth, but predicting the length of a downturn is extremely difficult. Therefore, the only appropriate business reaction to a market downturn is to react immediately. The objective is to profit during a downturn, which is impossible without downsizing. Many contractors argue that it is their business, and their choice to make. The response to that is similar to the choice a person has when facing a '*don't walk*' sign at the red light of a busy interaction. You can wait for the '*walk*' sign and green light, or you can walk out into traffic. A contractor may believe they have a choice not to downsize but a businessperson has no choice. They must downsize because risking even the potential for financial loss or failure is never an appropriate decision.

Capturing, training, and retaining good people has been a major issue for contractors during any growth market, so serious resistance to some of these *construction business realities* is understandable. Construction professionals, like all executives, must sometimes react to circumstances they can't control with processes they don't like. A safe or prudent reaction often requires valor. It is difficult but necessary to keep emotions out of business decisions. One would do well to remember that managing risks to the organization protects the jobs of the employees that remain after necessary cutbacks. Downsizing also provides an opportunity to weed out weaker employees. The objective is to profit from the reduced volume. It is entirely possible to survive during a down cycle. However, only if the company is able to minimize losses will it be able to achieve the objective of protecting its financial capacity. Financial capacity that will be critical when the rebound occurs.

Managing Growth

A study of the economic dynamics occurring during a construction market downturn or rebound indicates that both can present a financial struggle for contractors. The difficulty in good markets is that growth eats cash, and many firms have difficulty financing rapid or excessive growth. Research confirms that the failure rate during strong growth markets can be worse than during a downturn. Construction organizations need to manage cash flow judiciously during growth periods to remain financially viable and creditworthy. They need effective strategies to deal with the significant cash flow demands of growth and may require discipline to resist the *feeding frenzy* of loading up with too much work in rapidly expanding markets. Comprehensive strategies are necessary to manage growth risks and avoid losses during a market downturn that depletes capital and diminishes equity. These weaken a firm's ability to share in the recovery when it occurs.

Recovering Markets

In the longest of the construction market downturns occurring in the last seven decades, aggressive bidding persisted right up until the market returned to the size of the market prior to the decline. Contractors often persist in aggressive bidding following a downturn in an effort to get back to their original size. Continuing aggressive bidding during a recovering market compresses margins when they are needed most. Weak margins tend to lengthen recovery time and create a need for outside financing, which is problematic for firms financially weakened from a previous down market. Add to this the uncertainty about construction lending within the banking industry during extended recoveries, and it becomes the *perfect storm*. Prospering in cyclical markets and surviving both a recession and recovery in the construction industry starts with recognizing what is happening in the marketplace. In a cyclical industry market dynamics can fluctuate at any time for what may seem like no reason.

If for any reason there are fewer projects in any market, even temporarily, competition intensifies, and prices and profits tend to diminish quickly. In a shrinking market it would be ideal for each contractor to accept proportionately less work so that the market share of each individual business is maintained. However, there is a tendency in the construction industry to resist any reduction in sales, often strenuously, and to fight vigorously for the fewer available projects, driving down prices for everyone. This tendency persists, perhaps even stronger through initial and intermediate stages of recovery, when every project seems like it may be the last project to be seen for a while. Any attempt to increase market share is always and universally *bought* at a price. Trying to return to a prior size rapidly after a slowdown has the effect of limiting profit potential and magnifying risk in an already risky period.

This exposure is compounded by the fact that in each stage of recovery, particularly after a sustained downturn, cost inflation is common. It is caused by labor shortages and material cost escalation due to depleted inventories and production capabilities. Some laid-off construction tradespeople and managers who have departed from the industry do not return or return only after they have confidence in a sustained recovery. Material manufacturers and suppliers have suffered the same financial difficulties as contractors, having been forced to cut back capacity and are unable to raise prices. Regaining capacity requires investment and takes time, causing material shortages that lead to rapid price increases. If history repeats itself, contractors attempting to make up for lost ground after a downturn will load up on cheap work, increasing their risk in already difficult circumstances. You cannot control the market, but you can control your response to it.

Mobility vs Margins

Decades ago, construction was a local activity. Contractors did not often come from another city or state to bid on work in another city or state. When there was a lot of work locally, all area contractors also reaped the benefits if they downsized voluntarily. When there wasn't a lot of work in an area, contractors in that area still made money, just less of it. Today contractors come from all over the country (even all over the world) to bid on work. As a result, margins and markups on all work have been reduced. The reason is that margins are basically impacted by two things—the available amount of work and the available number of bidders. Historically, in less-mobile eras with a fixed number of bidders, if there was a lot of work, margin went up. When there was less work and a fixed number of bidders, margin was maintained or slipped slightly. Today with an unlimited number of bidders, prices stay down. If contractors had fixed territories, margins would stay up, but that is not likely to happen, so contractors need to work with the new realities. This unfortunately means continuing to obtain work with low margins.

Technology

Prosperous nations are going through a continuous *technological revolution*. Construction was one of the last industries to embrace new technology, but now it is coming towards us faster than anyone can react. Few can make full use of new technology before it is replaced with a newer improvements of the technology. Purchasing new technology is expensive, and introduction and training add to the cost, so small and mid-size contractors are hesitant to invest in new technology. The biggest technological risk is that the first iteration of a new technology may be improved so fast that it becomes obsolete before it pays for itself. Industry margins can't support trial-and-error attempts at technology. Poor selection, or just bad luck, can be costly. Expensive technological mistakes or missteps can be fatal. Some contractors wonder if it is worth the effort, and only large companies seem to be willing to take the risks. Despite current labor shortages, new technologies that offer labor savings and quality improvements are slow to be embraced by the construction industry.

Debt

Construction is the only major industry that puts the work in place at their own expense, invoices for the work after the fact, and is then paid only 90% of their costs incurred. Retainage of 10% is commonly held until completion of the project and final payment is made. In this industry final payment could be months or longer after beneficial use of the project by the owner. Retainage is supposed to protect the project owner from contractor non-performance, and to ensure that all project costs are paid. This protection is on top of the requirement of payment and performance bonds in the amount of 100% of the project price guaranteeing the same thing as retainage. Between retainage and, almost standard late payments from project owners, contractors are, in effect, financing their own work. With payments often 60 or more days late, contractors are also providing free project financing to the project owners. It is little wonder most, if not all, contractors of any size or type are in debt paying interest for what seems like just the right to be in business.

The amount of debt held by contractors is not sustainable. In a stable industry the debt-to-sales ratio should be constant. If a business with ten million annual sales owes one million dollars, it is fair to assume that when that business grows to twenty million annual sales it will owe two million dollars. In a 20-year research project, it was determined that the debt-to-sales ratios of a statistically significant number of successful equipment-intensive heavy highway contractors was increasing continuously. Over the 20 years there was an order of magnitude increase in this ratio that should have remained stable. To verify that there is no reason for this to occur, see the chapter on equipment cost. This trend will threaten the stability of the industry, and if it continues, it will have the potential to lead to crisis proportions.

Declining Market

While it takes time to become good planners and to forecast effectively, it is important when predicting the market to guard against optimism and to accept a downturn if that is what the data indicates. There is the potential for success or failure in the construction business during both growing and declining markets, but only if you know how to right-size your organization. The options are to fight for every job by reducing margins or to reduce overhead and prepare for the market decline. In downsizing you keep the best people, which increases efficiency and productivity. Contractors and top management get closer to the work when layers of people are reduced. Most managers work long days already, so care needs to be taken that work times are allocated appropriately. Fortunately, downsizing generates cash because money comes in from the previous greater amount of prior work, and spending is less for the current lesser amount of work.

When you see a declining market coming, you need to decide what reduced level of sales will be acceptable without reducing prices too much. This is a difficult calculation. The less work you are trying to capture in a declining market, the more likely it becomes that you may be able to maintain margins, or have to reduce margins less than if you are trying to capture a lot of work. All of your competitors will be after the same work, so it is a matter of desperation. Some will strive for no reduction in sales, which can lead to below-cost pricing. That means the share of any profitable work will be greatly reduced. Understanding this is critical because it puts into perspective that if you want projects with some profit on it, then you may have to seriously downsize. If you can't accept that then you will have to join the desperation pricing and risk losing everything. Many have tried to find a happy medium or compromise position, but there are none. A declining market presents huge challenges, which is why it also causes a huge number of construction business failures. And many of these failures occur years after the market recovery. The thing to remember is that there is always a market recovery. Your job is to preserve your organization so that it can benefit from the eventual growth market.

The question is, how can an organization find shelter in a declining market? It is clearly safer to downsize than to take work at breakeven or worse. Downsizing is simply a deliberate shrinkage of a company's overhead in anticipation of a decline in sales volume. It is a profitable alternative in a market downturn. However, in an industry where bigger always means better, downsizing is loaded with negative connotations. The steps are difficult and can be painful, but many who refuse to downsize face serious risk of losses or worse. It is, however, an intelligent response to a tough economy. Downsizing requires an immediate reduction in overhead to fit the new and reduced volume. Because overhead is primarily employee and equipment costs, the difficulty lies in determining who has to go and what equipment needs to be sold. Putting equipment up on blocks doesn't work well because of continuing insurance and depreciation costs. If or when the firm returns to its pre-downsize sales, it may or may not decide to return to its prior permanent overhead. The alternative is *Flexible Overhead.*

Flexible Overhead

Flexible overhead is a brand-new concept for contractors who want to be successful in good markets and bad (Figure 16.2). The old paradigm of continuous growth doesn't work in a cyclical market, which means that the contractor of the future needs to be organized to go up and down in annual sales to cope with clinical market realities. The construction market is cyclical period. This means that every 10 or 15 years there will be a market downturn. This, combined with the reality that a construction professional should never be inclined to chase inappropriate work just to maintain sales means there is no choice other than to be prepared to downsize on short notice. The old growth model only works in a growing, healthy market, and it accounts for the industry-wide drop in margins and excessive failures in a declining market when contractors refuse to give up hard-earned growth earned during better times. The reality is that construction has been a cyclical industry since WWII and the cycles will continue. The drive for size must be substituted with a drive for profits.

An overhead paradox that we are stuck with is that half a person, half a truck or half a piece of equipment can't be hired or purchased, but often that is what is needed. To guarantee profitability in a declining market, contractors need to be able to reduce overhead expenses by 15% to 25% within a short period of time. Preferably within a week or less. While skeptics suggest this concept is impractical or expensive, it is a proven model that can be easily accomplished, but it requires an alteration in business practice. During down markets too many construction organizations give back in losses what they earned during profitable years. Many years of research and testing have established that flexible overhead is like low-cost insurance against losses incurred during declining construction markets. When work falls off and overhead can't be reduced quickly or at all, losses result. Flexible overhead prevents these losses, and maximizes profits during all market conditions.

Figure 16.2 Flexible overhead allows contractors to scale quickly with market cycles, preserving profitability in both upturns and downturns.

Because we know that the construction market in the United States is cyclical, it becomes obvious that the successful contracting firm of the future must have the ability to vary its workload. This means the ability to be profitable doing less work in one year and more work in another year. And perhaps even more work in the year after that. You also need to be able to make these adjustments rapidly and seamlessly on short notice. The successful construction company in the future will be driven by the market, instead of being driven by a demand for a steady stream of work. Something that is impossible in a cyclical market. In the past, growth meant permanent increases in overhead because almost all overhead expenses are permanent costs. Overhead expenses are also much easier to put in place than to get rid of. In a cyclical market permanent overhead increases become a burden and create losses almost automatically when the market slows.

In the past, a common reaction in a declining construction market was to search for work in unfamiliar markets, such as bidding work in new geographic areas or for different types of projects. It always seemed like that was the only solution to hold the organization together. And most often that strategy led to reduced profits or losses. A declining market obviously presents risks for a construction organization, but moving into unfamiliar markets or attempting new types of work just magnifies that risk. It is impossible to combat risk by taking on more risk. Market slowdowns can and do occur during good markets. For example, when an organization's workload falls short after missing a few consecutive projects in a row; or unexpected delays on project starts. Too often contractors are tempted to underprice work or look in other markets just to maintain sales volume rather than cut overhead. Construction firms claim they have no choice but to do these things, but the reality is that there is an alternative. The choice is to accept a reduction in sales and *adapt* to the current market realities.

A far superior and significantly less risky strategy when available work declines and sales drop off is to reduce overhead in order to maintain continuous profitability. This may seem impossible and a huge paradigm shift from the often-heard standard of: *If you're not growing, you're going backwards*, or, *I have to keep my organization together for when the market comes back*. My response to these claims is to by ask: *At what cost*? Lengthy and painful down markets in the past have demonstrated that excess overhead costs weakened numerous construction companies. It took many years to discover an effective practice to counter overhead challenges during market cycles. The discovery was *Flexible Overhead*, detailed in the following section.

The Practice of Flexible Overhead

The answer to overhead management during downcycles is a strategy the author calls *Flexible Overhead*. This overhead structure allows a construction company to vary its sales and remain profitable. Hundreds of firms have adapted *flexible overhead* and are able to downsize when necessary and still remain profitable while their competitors are losing money because they have no control over their overhead. The process entails the conversion of a portion of permanent overhead costs into costs that can be discontinued in a week or less. In some cases, costs can be reduced in one day. To accomplish this, a select portion of overhead costs were converted to non-permanent expenses such as temporary personnel, rental equipment, interim office space, and more. The percentage of overhead costs that are coveted varies from 15% to 25% of total overhead expenses. The variation in the percentage of converted overhead is because organizations vary in their overhead structure.

There will be skepticism with some suggesting these steps may increase expenses or fear that temporary personnel are less qualified. Many claim it is cheaper to own equipment than to rent and so on. The response is that the cyclical nature of the construction market renders continuous growth in sales impractical as a sustainable business model for construction firms—particularly in an already high-risk industry. A focus on growth in profits rather than sales is the only appropriate objective. Maintaining size at any cost in a cyclical market can lead to economic suicide. However, when growth in profits becomes the focus, the selection of projects shifts from a continuous need for sales to selecting projects that the organization does best. With flexible overhead team members have job security because the inherent stability protects the company and its staff.

A concentration on profitability and not size presents a different understanding of the construction market and the firm's place in it. This does not prevent entering new markets, but it does tend to cause a more careful measurement of the associated risks with a heightened appreciation for the potential learning curve and an honest appraisal of the term's abilities. The adoption of *flexible overhead* also tends to foster a new definition of success that includes careful measurement of risks. This is not to say that revenue growth is never advantageous, but it only makes sense if the market is growing and believed to continue to grow. There is a time for revenue growth, but it must coincide with maintaining profit margins. Growth almost always sacrifices some productivity and efficiency at least temporarily which negatively impacts profit margins. Growth is a strategy that needs to be managed very closely with an eye on profit and an understanding of the sacrifices and inefficiencies growth causes. Growth at any cost is out of the question, as is growth without a detailed plan. It always makes more sense to focus on profit, which remains the prime objective. Being selective in the choice of work, and understanding the company's core competences, can lead to organic growth without sacrificing profit.

Construction companies have traditionally held to a belief that they have to keep their entire organization together during slow periods because it took so long to find and train their staff and get them to work as a team. Some contractors believe it would be impossible to replace those workers when the market rebounds. During a lengthy market downturn, almost all construction organizations have been in the position of eventually having to cut overhead. However, the real question is whether they should have made the cuts earlier. On the other hand, during market recovery, the challenge is to determine what overhead to put back into place, in what order, and when. Should the overhead have been added in the first place? Did it enhance efficiency and profitability? Or was it just a way of maintaining a comfort zone or safety margin?

Flexible overhead deals with the reality of a cyclical market and prevents the desperate need for sales to achieve an arbitrary sales goal. To be flexible, a construction company needs to rent some portion of their equipment so that it can be returned when work inevitably slows down. Similarly, employing temporary office and accounting staff assures a portion of costs that can be discontinued with short notice. Even if these options are more expensive compared to the alternative, the modest differential in cost buys flexibility. It can be considered survival insurance. During a slow market, it is not the time panic. It may be appropriate to look for smaller projects closer to the company's base of operations to allow thorough supervision to increase efficiency and profits on the lesser amount of work. Leadership is growing and enhancing *people*, not just sales. A little less volume may provide time for management to increase efficiency and design an improved overhead structure for when the market recovers.

A study of overhead reductions during the 2008–2013 construction market down cycle identified which overhead expenses firms chose to reduce. The respondents represented companies of various types and sizes, and most made no overhead cuts for at least a year into the slowdown. After waiting too long, nine out of ten companies cut some overhead—70% reduced bonuses, company functions, and charitable donations. Forty-one percent reduced business development expenses. One out of five contractors cut between a quarter and half of their overhead, and 15% cut overhead by more than 75%. Contractors reported that they cut corporate salaries, training, education, and travel along with large cuts in retirement plans.

Managing Overhead

General and administrative expenses (overhead) are obviously necessary for the efficient management of an organization. For a cyclical industry like construction in which sales often fluctuate considerably, the question is how much overhead is necessary or appropriate relative to sales volume. This number may differ each year and sometimes each month. The largest portion of overhead expenses is made up of salary and salary-related costs, and these costs are generally considered permanent positions necessary to the functioning of the organization. Small and mid-size closely held companies are often managed more like a family than a business. Their approach is noticeable when it comes to bonuses and perks, especially during good years when profits often are shared more generously in closely held firms than in larger or publicly traded enterprises.

When the market slows, the reality is that it is much easier to put overhead in place than it is to reduce it. Many closely held enterprises simply accept the losses created by too much overhead during a market slowdown as a cost of doing business. All contractors need to manage overhead more carefully to eliminate the decline in employee morale that is caused by cutting excessive overhead during a market slowdown. Flexible overhead facilitates rapid overhead cost reductions during volume fluctuations eliminating the need to fill volume gaps with low-profit or no-profit work. The evidence is clear that cutting overhead as soon as the market turns down is the appropriate approach. In addition, every category of overhead expense should be scrutinized regularly, and adjusted at least annually in anticipation of the following year's sales. Decreases should not be made after market downturns, but rather in anticipation of downturns. To prevent financial losses and remain profitable, contractors need to plan ahead.

It has been standard industry practice to embrace permanent overhead in preparation of becoming the new larger company that we intend to be. However, this makes no sense in cyclical market. To be successful in a cyclical market, an organization needs to be structured to be compatible with a fluctuating market. That means being able to maintain profitability by doing less work in one year and more work in another year. Successful contractors will be driven by the marketplace. Flexible overhead is the only way to deal with the realities of a cyclical market. To be flexible, a contractor will rent or lease some portion of their equipment in a manner that the costs disappear when the work slows down. Utilizing some portion of temporary office and accounting staff that can be reduced on short notice ensures flexibility. The modest differential in cost is meaningless compared to the assurance of continuous profits. Call it *survival insurance*. Ground rules include the avoidance of jobs and positions that don't fit into the plan and never embracing work that interferes with flexibility.

Summary

Whether anyone knows it or not or believes it or not, the market is cyclical. Successful construction professionals will embrace this reality and manage accordingly. A cyclical market cannot and will not support a continuous growth strategy, which means having 100% permanent overhead needs to go. Too much overhead has caused some construction companies to become office-centered to the point that they have little time for the workforce. Some firms have too many white-collar employees who do not build anything. These need a few less desks, telephones, and keyboards. It is possible to have too many people collecting data that no one uses and adds nothing to production. Some need to get back to basics and back to when the focus was on supporting the field to perform the work at a profit. Absentee leadership cannot control the office or the field. Leadership is growing and enhancing people, not sales. In a down market you have a short-term problem that can be managed if you are willing to do a little less volume for a while. The sooner the industry embraces flexible overhead, the sooner the failure rate will finally begin to reduce. The choice is ours.

17 The Value of Seeking Advice

Introduction

The construction industry, with few barriers to entry, is one of the few major US industries with more small- and mid-size businesses than large and massive-size firms. The contractors in small and mid-size organizations are, for the most part, hands-on leaders. As such, they spend the majority of their time building the work or causing the work to be built rather than managing the business. The initial and primary cause of the success of these firms depends on the contractor's ability to put the work in place for a profit. For a startup contactor, that ability is untested. Weaknesses in that ability are likely the main reason for the greater than 50% failure rate of startup construction companies within the first ten years. Failed startups never actually become a *business,* as such. Those that succeed longer than ten years can rightly claim the designation "a business". They should also understand that they have joined an industry with the second-highest failure rate in the nation. If all these successful firms have proven their ability to build for a profit, then what are the causes of the industry having the high failure rate?

Let's explore that question. There are two distinct sides to every construction operation: the technical or performance side and the administrative or business side. If all these organizations have proven that they can *perform* for a profit, the high industry failure rate must be a result of occurrences on the business side of their operation. This will not be much of a surprise if you consider the experience and expertise of these *successful* construction professionals. They either learned from experience in the field or were educated in engineering or construction management. Neither of these learning tracks includes any advanced courses in accounting, economics, business law, or management science. Herein lies the problem. Knowledge of these subjects is critical to managing the business side of any for-profit company, which obviously includes construction firms. Neither construction nor engineering school curriculums include advanced courses in accounting, economics, business law, or management science. Experience working in the field obviously offers no exposure to business courses.

Some construction professionals claim that they learn these subjects simply by being in business. Others claim that entrepreneurs have a *natural intuition* about these topics. Neither of these explanations makes any sense at all. Some argue that the technical side of a construction business is so much more important than the business side that the business topics don't matter that much. A few have claimed that little time needs to be spent managing the business issues in a construction company because technical proficiency is all that matters. Both of these positions defy common sense. The reality is that successful contractors with limited business education or experience underperform or fail without a reliable source of critically necessary business advice, advice that includes accounting, economics, business law, and management science to remain successful. The most appropriate and reliable source

DOI: 10.1201/9781003500315-18

of this advice is the firm's qualified Chief Financial Officer (CFO). Construction organizations with strong CFOs seldom fail. This reality is validated by the author's years of research into the common causes of construction business failures. Unfortunately, as discussed in previous chapters, the input or influence of CFOs on the management of the organization is sadly lacking in most construction firms.

Separate Roles

Because almost all small and mid-size construction firms, and many large construction firms, are owned and operated by the same person or persons, it is appropriate at this point to explain the different skills required for success. Business owners and the managers of these businesses have separate and distinct roles, and when they are the same person, one might say *wearing two different hats*, it can be confusing. The issue is that while you may own and manage a business, you can only wear one of those hats at any one time. It is critical to understand the separate roles of owner and manager in the closely held business. Family and closely held businesses have unique issues that do not affect businesses in which the owner and top management are separate; or where top management does not own a controlling interest in the firm, as is the case with publicly traded companies.

The roles of the owner of a business and the manager of that business are separate and distinct, and each role is equally significant to the success of the enterprise. This is complicated by the fact that in small and mid-size businesses, the same person is usually responsible for both roles. This means that in the closely held construction company, the contractor wears both owner and manager hats. Unfortunately, few entrepreneurs seem to understand that they cannot function well wearing both hats at the same time, because these are distinctly separate roles. Roles that on occasion will have conflicting requirements.

The roles of owner and manager are so different to the point that critical decisions by a manager for short-term gain are often in opposition with the long-term interests of the owner. As a result, many closely held businesses suffer from a lack of clearly defined objectives and long-range strategies. The future of a business can vary dramatically depending on which direction a single management decision happens to take the firm. It is the responsibility of the business owner to establish long- and short-term goals and objectives for the company and to discipline the performance of management toward the attainment of those goals. However, when both management and owner decisions are made by the same person, a critical analysis of the impact of each decision on corporate goals is often overlooked. Many construction company owners believe that as long as everyone is working hard, putting in the long hours, and seeming to care about the success of the business, everything is fine. Loosely translated, this means the measure of success is *that everyone is working hard, putting in long hours, and cares*. This of course is not the case.

Management is supposed to organize the resources of the organization to achieve the short- and long-term goals and objectives set out by the owners or stockholders of the company. If that is not accomplished, management is also expected to be prepared to explain in detail why. When the owner and management are the same person, realistic goal setting is often supplanted with the intent; *We'll push as hard as we can and then measure the effort later*. When the effort is measured later the results are usually; *Where we are at is where we should be*. Many owners of closely held businesses refuse to measure their own performance, contending that it is nobody's business how they run their company. They fail to realize that a fundamental definition of running a business includes *setting objectives and measuring the performance of those charged with attaining those objectives*. To do anything less is to invite disaster by just continuing to work hard until something goes wrong by allowing a business to grow until it self-destructs from lack of direction.

Owners Direct

When the differences between the roles of the owner and manager are recognized along with the reality that no one can wear both hats at the same time, the rest is easy. Owner/managers of closely held businesses need to set aside time at least three or four times a year to take off their *management hat* and put on their *owner hat* in order to perform the important functions of their dule roles. While wearing their owner hat their role includes:

- Establishing and updating the short- and long-term goals and objectives of the organization
- Reviewing and recommending business policies and procedures for the enterprise
- Objectively measuring the performance of management (that includes themselves) in following the organization's policies and procedures and accomplishing the company's objectives

A general rule of thumb is that business owners do not necessarily concern themselves in the day-to-day issues of running the business. In small businesses it is helpful, but not easy, to avoid mixing owner and management issues. It helps if the owner's issues can be isolated in some way, for example, dealt with away from the office, after hours, or when no interruptions are likely. This also helps facilitate objectivity; and where there are multiple owners it helps neutralize any rank and title difference of the owners. For sole owners an alternative to making all critical owner decisions alone is to form a small group of knowledgeable trusted employees to assist in the process. Some call it a leadership or management committee that assists the owner in determining what is right for the company, setting attainable goals, and developing ways to measure performance. This committee usually includes top managers, but some contractors get even better results when middle or field management is represented. Some even include one or more trusted employees from the workforce.

Many enlightened owners are beginning to include labor and first-line supervisors in business planning with exciting results. The people who do the work often know a lot more about how the operations really function than they are given credit for, such as:

- The strengths and weaknesses of manpower resources
- The overall potential of the field organization
- The quality of technology in place
- What's right for the company

These internal employee committees can assist in clarifying what is best for the company. They can function in this role completely independent from their day-to-day roles within the organization. Establishing even fundamental corporate goals has had profound effects on both the owners and employees of small closely held businesses. Participation gets people excited about what is right for the company and how to achieve it.

Seeking Advice

This is genuinely a controversial subject. It is hard to advise self-made successful contractors that they don't know everything. Some consider it insulting. Success builds self-confidence. However, if you have built your company from the ground up to a meaningful level of success, why seek advice? Self-confidence is comforting and positive, but it also exists very close to an invisible line called *over-confidence.* This is not a challenge. It is a fact backed up by the reality that no one has ever self-reported being *over-confident.* This should be of interest to all of us because we also know that it is very common to hear people being accused of being *over-confident.* Apparently, anyone who is overconfident would ever be aware that they are overconfident. Which brings up the question; does *over-confidence* even exist? If *over-confidence* exists at all, it is clearly NEVER recognized by anyone who has it. Therein lies the danger. Overconfidence is always dangerous because no one ever knows if they have it. For a business owner, it is lethal. As a compromise, let's just agree that no one knows everything about everything. Smart people test some of their beliefs with others. Brilliant people test all of their beliefs with others. They seek out people smarter than themselves and discuss and test their ideas before implementing them.

Board of Directors

The best approach for even the smallest contractor to position their company for success is to establish an active and independent board of directors. A board of directors assumes the owners' role as the representatives of the stockholders. Some prefer the less formal term board of advisers, but I consider them interchangeable because these advisers' function at the will of the owner or stockholders, so the contractor remains in ultimate control. Most closely held businesses have an inactive board listed in their corporate documents and some have a few family members or friends named as board members, but neither are typically active nor provide *independent* advice.

Boards made up of *insiders*, that is family members or close associates like the firm's accountant, lawyer, or banker, are too close to the owner to be objective enough to measure and discipline management's performance. An effective board of directors or advisors has a majority of truly independent outsiders. For contractors who are not ready for that much *outside help*, as few as two or more genuinely independent outside board members can provide extreme advantage. I can't express strongly enough how much value small contractors have received from the advice and recommendations of genuinely independent board members. However, contractors tell me regularly that they don't need anyone second guessing them and my response is always a declaration of my fear that a lack of checks and balances on strategic business decisions may be the primary cause of construction contractors having the second highest failure rate in the nation.

Forming a Board of Directors/Advisers

Some small or mid-size business owners shy away from a board of directors because they believe there are too many legal considerations to contend with. While this may or may not be a substantive issue, some prefer the term *board of advisers*, which they can define in any way they choose and does not imply any legal obligations. Either works for the purpose intended here because seeking advice is not recommended for legal considerations but for practical purposes.

Board members should be owners or senior managers of successful businesses who have personally had profit and loss responsibilities. Preferably they have been responsible for "making payroll" at some point in their careers. Making payroll means they were responsible for seeing that there was enough cash flow to support all ongoing cash needs which is a test I use to define an "entrepreneur". The size of the candidate's business is not critical. However, if their company is much larger than yours, they may not consider their board role as important as you do. The owner of a growing printing business or successful auto dealer may be an ideal candidate. Bankers with whom you are not doing business make good board members. If you are weak in marketing, an advertising or public relations executive is a good addition to your board. I do not recommend attorneys, accountants or medical doctors for board positions. Their experience is with their "practice" which is not a commercial business. They therefore lack what I call actual business experience or operating business experience.

Board members are compensated for their efforts, and the amount generally depends on the size of your company. Compensation establishes the professional nature of the relationship. Board of Director fees for small firms vary considerably with location and size of company. By way of example, a board fee for a small construction firm up to about 10 million dollars in annual sales, may be as little as a thousand dollars or more per meeting. Fees vary from rural areas to big cities and for larger companies, board fees increase considerably. Some large companies pay annual board fees in addition to an amount per meeting. Making inquiries among industry professionals should clarify current local or regional fee structures.

A board should meet no less than two times per year, or the members become too far removed from the business to be effective. Boards should meet no more than six times per year, or the members tend to become *insiders* and lose objectivity. Three to four times per year is effective, unless there are numerous changes going on within the organization or its environment. The number of board members is optional. For small and mid-size companies five to seven members is common. Some prefer to start with fewer members until they see how it works for them. Three members is a minimum and can be effective as long as two are independent board members, which means they are not employees or service providers to the firm. I do not recommend family members on boards but realize that sometimes that cannot be avoided. While an odd number of board members is standard, there are more discussions than voting on boards of closely held companies, so an odd number is not imperative.

An active, independent board of directors may be the single most important choice the owners of a closely held or family business can make in ensuring their future prosperity. As a contractor you already know how to produce the work. What you need from the board is business advice. Information, clarifications, and recommendations on how to manage a business. Boards assist with organizational issues, business planning, recognizing and managing business risks, and much more. They help contractors manage their *two-hat* responsibilities. Boards also influence a contractor's responsibility to manage their ownership role objectively and separately from their heavy management responsibilities. This is vital to the leadership and success of a closely held business (Figure 17.1).

Figure 17.1 Choosing a board of directors can be crucial for the owners of a closely held or family business.

Case Study: Board of Directors

Within a couple of years of creating my consulting firm to provide failed project completion services to sureties, my company was overwhelmed with new business. Located in the Northeast United States, I had already opened a branch office to serve the Southeast and was still overwhelmed by the opportunity for further expansion. I could manage the services my firm offered but realized I had no idea how to operate a national consulting organization or how to expand it across the country. Fortunately, I realized I was in-over-my-head and needed help with deciding what to do and how to do it. My senior executives were well-educated engineers, but none of them had ever owned a business and neither had I.

I needed a board of directors. Fortunately, I was familiar with a number of small business boards made up of family and friends and often the firm's lawyer and accountant, and my experience was that in every one of these circumstances the boards were ineffective. The board members were insiders and non-business owners, so they had no more experience in running an operating business than the business owners whom they were supposed to reinforce. I realized that I needed experienced businesspeople as board members. I also realized that my board members did not need experience in construction because I did not need technical construction advice. I needed business management advice.

Once I understood that, the selection of outside board members was not difficult. I looked for people who have been successful in business. It was apparent to me that they did not need to know anything about my type of business. Not knowing anything about construction was actually preferred because the role of the board is not to participate in the running of the business. Board members expect management to do that, operate the business. It is not hard to find successful people in the local business community who are usually flattered to be considered as board members. I asked my managers, associates, and friends who they might recommend for a board position. I also knew there was no need for me to be acquainted with any perspective board member and that it may be preferable if I was not. I interviewed candidates without any commitment in advance. I was looking for people that I could trust and that I would be comfortable working with.

The results were amazing. My board of five members including myself and my brother who was not active in my business and three senior business executives that I had never met before. One member was president of a national wholesale food production company. Another was the executive vice president of our statewide utility company. The third was the president of a bank in another county that I did not do business with. His bank required a written agreement that I would not do business with that bank in the future. Prior to interviewing these senior business executives, I had never met two of them and was only acquainted with one socially.

The opportunities for my consulting services were clearly national, but I had no experience or conception of how to accomplish that level of expansion. At the second quarterly meeting, the board asked that a study be made of how many regional offices it would take to provide our services to the surety industry nationwide. At the third quarterly meeting I presented a national map showing potential offices in Dallas, Texas; Chicago, Illinois; and San Francisco, California in addition to the New Jersey and Florida offices. I suggested to the board that some years in the future perhaps we could open an office in Dallas, and then some years later in Chicago. If they were successful, we might then be able to attempt an office in San Francisco.

The board meeting lasted all day, and the board decided we should immediately open an office in San Francisco so we could describe ourselves as a national firm. They said that six months later we should open offices in both Dallas and Chicago at the same time. Needless to say, I was overwhelmed, but the board felt that the timing was critical. The board went so far as to describe how to conduct the search for regional vice presidents, how to use an industrial psychologist to screen them, and how to fund the expansion. One year later all the offices were open, we had over a hundred consultants on the payroll, and almost as many support staff. By then we had two corporate aircraft, and I still could not keep up with the new business coming to us. A year after that, in response to the demand, we opened an international division.

This expansion had very little to do with me. I knew how to provide the consulting and construction services our clients needed. I knew nothing about market studies, opening new offices, or organizing a huge organization. (*My guess is that most of my readers do not either.*) However, my board members did have that knowledge and experience. These issues were second nature to them, and the solutions and processes to accomplish these tasks were not a challenge to them collectively. I cannot honestly take credit for the success of my firm. The only credit I could take was knowing what I did not know. And knowing enough to find board members that knew more about business issues than I did. It also helps to motivate them and then listen to them.

Internal Decision Making

If the above discussion about decision making makes any sense to you at all, then this is the perfect opportunity to address internal decision making. The field operations of the typical construction company do not enjoy the benefits of consistency and repetition that, for example, the manufacturing industry has. Manufacturing is conducted in a controlled environment, typically indoors, with highly repetitive processes. Whereas in the construction industry site-conditions vary, and most of the work is conducted outdoors in varying weather conditions. In addition, specified materials differ from project to project, and the individuals performing the work vary with each project and sometimes from day to day.

Suffice to say there are more inconsistencies from project to project than consistencies even with very similar projects. Differing site conditions and variations in materials and designs require a huge number of small to large decisions to be made daily. For clarifications and major decisions, a worker may ask a foreman, a foreman may ask a superintendent and so on up the chain. But what about the numerous minor decisions tradespeople must make constantly such as: *Should I put one temporary brace on this or two? Do I put one nail in this or more? Should I dump this here or there?* These decisions can be consequential and can have serious ramifications regarding quality, safety and productivity.

Safety experts who have investigated construction site accidents suggest that most small and large accidents were caused by minor errors in process or procedures. That means that the negative results were obviously preventable. The same can be said about the causes of the need for rework that impacts productivity and profits. Construction field operations have historically been employing top-down management and supervision. Or to put it another way: *Do it the way I told you.* This means that workers need answers from their supervisors. The construction industry has been slow to advance its safety procedures compared to the manufacturing industry which has been listening to and involving the workforce in process improvement, reporting problems, continuous inspection, and safe working conditions for decades. Those closest to the face of the work are often the first to uncover problems or recommend improvements. However, a strict top-down management approach, as found in construction, does not necessarily listen to or encourage worker participation. Many years ago, when I joined the industry as an apprentice, worker suggestions or speaking out was actually discouraged.

It is passed time for the construction industry to embrace the fundamentals of process improvement by including the workforce in determining work processes, procedures, and sequencing. The construction industry needs to follow the example of the manufacturing industry. In manufacturing today, recommendations from workers are encouraged and even rewarded. This process has improved manufacturing productivity, quality, and safety, resulting in increased profits. Some enlightened construction firms encourage worker participation, have empowered workers to speak up, make recommendations, and take initiatives. This has resulted in much improved project safety and productivity, and in most cases a marked improvement in worker satisfaction. It is happening slowly, but the construction industry needs to embrace the reality that the workforce collectively knows what they are doing and needs to be motivated to speak up. Their input and participation in safety and process improvement have produced amazing results in organizations that encourage worker involvement in the planning and production of the work.

Worker Continuing Education

Encouraging worker participation has also led to a realization that continuing education of the workforce improves worker satisfaction and pays dividends in increased productivity. Enlightened contractors have realized that a better-informed worker is a more productive worker. They have noted that on-the-job safety talks have improved project safety and realize that on-the-job toolbox talks can and do improve productivity. Industry studies have proven that the workforce is stimulated to better production when they realize that their employer is paying attention to them. One of the best forms of worker recognition is the offering of continuing education. Some will argue that the employees will be working for someone else next year or that tradespeople are supposed to know their trade. The reality is that improvements in the efficiency of the workforce benefit the contractor, so contractors should be anxious to fund continuing education. The cost of toolbox talks is minimal compared to the value of just one worker becoming more productive for a few weeks on their current project.

Educating the workforce on company time at company expense is a win–win proposition. The construction industry workforce is aging, and some of the techniques workers learned when they entered their trade have been much improved over time. All of the work processes and tools have improved. The opportunities for productivity improvement in the construction industry are huge, and there is no valid argument against upgrading the workforce. There is also no argument that someone else should do it. If the employer reaps the benefit of the continuing education of the workforce, then they should be jumping at the opportunity to provide it. If educated workers improve productivity, imagine the greater advantage of educating the workers' foremen and superintendents.

Case Study: Continuing Education

I was engaged by a large southwest contractor that had been suffering from deteriorating job profits for several years with the prior year's profits close to break even. Overhead expenses were under control, so the profitability problem was clearly poor production in the field. The firm grew considerably over the last five years, more than doubling their sales. This meant that half of their foremen and superintendents were either new to the organization or had been promoted from within. Of the several changes we made over the next year, the biggest was a *leadership education* program for the firm's more than one hundred foremen and superintendents. A week of all-day classes on leadership and production was set up for 20 people at a time. All foremen, superintendents and a few potential foremen and superintendents graduated from these classes. All were offered additional reading material for six more months.

The results the following year were that profits tripled from their prior year's poor performance. The second year after the classes, their annual profits were slightly above the national average. Four years after the company-wide leadership education program, total year-end annual profits were in the top ten percentile of the national average for firms of their size and type.

This is an example of what the upgrading existing employees by continuing education can do.

Obviously, there were also other company-wide improvements made, but the field productivity efforts produced the biggest return. It was easy to measure the improvement in field performance because it is immediately reflected in a direct increase in gross profit.

Nationwide contractors complain about the quality, motivation, and lack of experience of the available workforce. These complaints vary from location to location, but both union and merit shop contractors need to understand that they are directly responsible for the performance of their workforce. No one else is going to it.

There is no "typical" tradesperson. It is interesting how much productivity varies from company to company working in similar conditions in the same area. The difference in profit percentage from the least profitable to the most profitable is a factor of three? How can one group of similar contractors earn twice as much profit as the lower group and another group earn three times as much? This makes no sense and is not found in any other industry in the country. This data is for contractors similar in type, size, and location so they all have access to the same materials and labor pool. The performance differences must be attributed to how they manage and perform their work. What this verifies is that underperforming construction organizations can absolutely do something about it.

Summary

Being in your own business has its rewards, but for the most part they are hard-earned. Some have complained that it is lonely at the top. This is especially true for the small and mid-size enterprises. In the small firm you may be the only senior manager and certainly the only business owner, so there are no actual *peers* that share your concerns. In the mid-size firm, there may be one or more senior managers, but they will know little about the burdens of business ownership unless they have been in their own business. The truth is that *it is lonely at the top*. A board of directors or advisors' changes that. Unfortunately, only a modest portion of small and mid-size construction firms have independent boards. However, those contractors with independent boards have demonstrated improved performance, profits, and perhaps as importantly, a comfort level with less worry. They report valuing their interactions with board members who get it, who tend to understand what they are going through.

The reason small and mid-size businesses really need a board, particularly contractors, is that these contractors are *construction experts*—not *business experts*. They all know how to build the work, but that does not mean that they are automatically experts in accounting, economics, business law, or management science. The biggest hurdle is not finding board members. The biggest hurdle is realizing the huge advantages that a board of directors or advisors brings to even the smallest enterprise. The closest analogy I can think of is:

Would you rather be stranded on an uninhabited island by yourself?

Or with three people who think exactly like yourself?

Or with three people with totally different skill sets than yourself?

People tend to select board members like they select friends. They prefer people who are similar to themselves, who have had comparable experiences, and who share matching outlooks. Similar people make good friends. But they bring nothing new to the evaluation, development, and navigation of a commercial corporate enterprise. The purpose of the board is to bring expanded knowledge and experience into the business to support you with the evaluation of the organizational structure and business potential of the company into the future. Too many self-made entrepreneurs are convinced that they are already doing that themselves. When, in reality, most busy successful entrepreneurs tend to evaluate where the business has progressed to since the last time they gave it any thought. Because they brought the business to where it is at, they become convinced that it is exactly where the business should be. After all, they put all their energy into steering the business to this stage, so a celebration is in order, and as such, there is really no need for outsiders, and there is no time for that anyway.

Boards of directors or advisors are the single most underrated tool in the construction industry. Boards are the single most important (critical) ingredient in the recipe for success of the small and mid-size contractor. It is the same for large contractors. All large contractors that have active outside boards know it. Large contractors that do not have independent boards will eventually regret it, and hopefully not on the way to bankruptcy court. The major point here is that no construction business is too small to benefit from a board of directors or advisors dominated by outsiders. Boards dominated by employees of the corporation, professionals who do business with the corporation, or friends of the owners of the corporation are of no value. They may actually do harm simply by the erroneous *appearance* of independence.

Included in this chapter is the value of continuing education of the workforce. It is perhaps the most underrated and most cost-effective profit generator to self-performing contractors. The question of who else should be providing the education is meaningless. It is low cost and instantly effective, which should speak for itself. The construction industry is behind the curve on understanding the value and necessity of continuing education as a tool for enhancing profits.

Index

For Product Safety Concerns and Information please contact our EU representative GPSR@taylorandfrancis.com Taylor & Francis Verlag GmbH, Kaufingerstraße 24, 80331 München, Germany